THE GREAT EXTINCTION

Books by Michael Allaby

ECO-ACTIVISTS

WHO WILL EAT

ECOLOGY

ROBOTS BEHIND THE PLOUGH
(with Floyd Allen)

THE SURVIVAL HANDBOOK
(with Marika Hanbury Tenison, John Seymour and Hugh Sharman)

INVENTING TOMORROW

HOME FARM
(with Colin Trudge)

DICTIONARY OF THE ENVIRONMENT

ANIMALS THAT HUNT

WILDLIFE OF NORTH AMERICA

WORLD FOOD RESOURCES

MAKING AND MANAGING A SMALL HOLDING

POLITICS OF SELF SUFFICIENCY
(with Peter Bunyard)

A YEAR IN THE LIFE OF A FIELD

CURIOUS CAT
(with Peter Crawford)

ANIMAL ARTISANS

CHANGING UPLANDS

Books by James Lovelock

GAIA

THE
GREAT
EXTINCTION

Michael Allaby
James Lovelock

Diagrams by Paul Weaver
Picture Research by Science Photo Library

Secker & Warburg
London

First published in England 1983 by
Martin Secker & Warburg Limited
54 Poland Street, London W1V 3DF

British Library Cataloguing in Publication Data

Allaby, Michael
  The great extinction.
  1. Dinosaurs
  I. Title   II. Lovelock, James
  567.9'1   QE862.D5

ISBN 0-436-01160-3

Printed in Great Britain at
The Camelot Press Limited, Southampton

Contents

List of Illustrations

List of Diagrams

Introduction

In the late 1970s papers began to appear in the scientific literature of the United States and Great Britain that shed light on one of the great scientific mysteries of modern times. This book is an account of the unraveling of that mystery.

At its more superficial level the mystery concerns the fate of the dinosaurs, the great reptiles which "ruled" (we shall qualify that word strongly) the world long ago and whose rediscovery has delighted children (and adults!) since news of them first appeared about a century and a half ago. This aspect of the story is superficial not because the dinosaurs are uninteresting or unimportant, but because in our excitement about them we may be tempted to forget that when they vanished from the world some 70 percent of all living species vanished too.

Not every large reptile is, or ever was, a dinosaur. The word describes a large and rather disparate group, comprising very many species, but nevertheless it is precise. The reptiles were the dominant form of animal life on land for many millions of years, and the dinosaurs represent the culmination of their evolution. Far from being slow, stupid, maladroit, and maladapted beasts and evolutionary jokes fit only for cartoon strips, many were fleet of foot, all were very highly evolved, and in their time each of them was adapted supremely well to the conditions in which it lived. Until about 65 million years ago the reptiles in general and the dinosaurs in particular flourished. Then most of the reptiles disappeared from Earth. As far as we know, all of the dinosaurs vanished, and, as we have mentioned, nearly three quarters of all the species alive at the time disappeared as well. That is the mystery. What happened to cause so great a catastrophe?

We believe we know.

The mystery has remained unsolved for a century and a half, ever since the first fossilized bones were identified as belonging to a large reptile of a species unlike any alive today. All that was certain, although even this has been disputed, was that beds of sedimentary rock bearing fossils of these and many other animals lie immediately below other beds—which are more recent because they lie at a higher level—that contain almost no fossils, and none of the dinosaurs. The rocks bearing the dinosaur fossils belong to the period paleontologists and geologists call the Cretaceous, those overlying them to the Tertiary. The boundary between them marks the end of one chapter in the history of our planet and the beginning of the next. Then, at a higher level still and also in rocks of more recent origin, fossils occur again, but they are fossils of different species, which lived in a world dominated by mammals. The extinction, and it was a mass extinction, was sudden.

Many explanations have been attempted. Some were more plausible than others and some were simply silly. First one interpretation of events would be preferred, then another, until eventually there were so many explanations that many scientists gave up entirely and relegated the event to the realms of the inexplicable.

So it remained until a new discovery was made. We have said that the bed of Cretaceous rocks lies immediately adjacent to those of the Tertiary, but this is not quite true. Between them, in many parts of the world, there is a thin layer of clay, and when samples of this clay were analyzed their chemical contents were found to be quite different from those of ordinary clays and of the rocks above and below them.

This is where our story begins, with the "boundary clays." It is a little like a detective story. A body is discovered—in this case many bodies, of course. All attempts to explain the cause of death fail. Then, quite independently—for the scientists involved were examining something quite different and the coincidence in the dating occurred to them later—clues emerge. As the meaning of the clues is deciphered, a picture begins to emerge of a major catastrophe.

In trying to evaluate the clues, we believe we are breaking some new ground. Of course, much of our description of the event itself and its aftermath is pure speculation, but it is speculation that makes scientific sense. We may not describe what in fact happened, but we are sure we describe what *could* have happened. This is not easy, for the catastrophe itself is so far beyond anything that our species has experienced, so far removed from phenomena that have been produced experimentally, that it comes close to defying comprehension.

The catastrophe was immense—far, far greater than anything modern man could produce, even if every nuclear weapon in the global armory were to be released simultaneously. This fact has implications, and it is here that an ancient story becomes highly relevant. If the greatest disaster of which we have hard evidence failed to destroy life on our planet, how seriously should we regard the supposed threats that many people believe are posed by industry or warfare today? Could we destroy our own species, let alone all life? If we could not achieve even that, just how resilient is life itself? What would it take to change Earth into a world of only fossils, ruins, and ghosts?

Such are the afterthoughts with which we will leave you. First there comes the story, and that begins in the south of England with a woman. Her name was Mary Anning . . .

Michael Allaby,
Wadebridge, Cornwall

James Lovelock,
Launceston, Cornwall

1

A Body Has Been Found

Mary Anning (1799–1847) was no ordinary young woman. She had received little formal education, but she had an eye for business, and an unusual business it was. She found and sold fossils. Eventually she was to become renowned for her ability to locate specimens—the King of Saxony was one of her customers—and as her work prospered she learned a great deal about her curious wares.

She lived in Lyme Regis, Dorset, on the south coast of England, where the chalky rocks are rich in the remains of plants and animals that lived long ago and that, in death, were preserved and buried by sediments brought by the sea. Where the modern sea has eaten away at the face of the cliff, removing layers of material that arrived later, older rocks are exposed. It was along these cliffs that Mary searched, examining the contents of the frequent landslides and the chalk faces that were laid bare each time more loose material fell to the beach.

Mary's father had introduced her to the collecting of fossils. He was a carpenter by profession, but like many Europeans who lived in the early part of the nineteenth century he had a lively interest in the natural world and was a keen amateur collector of curiosities. Mary, who accompanied him on his walks, started her own fossil collection. Not everyone can live beside a chalk cliff and a market in other parts of England for fossils soon developed; Mary was only eleven years old when she sold her first specimen.

When Mr. Anning died and Mary was left to find a way to support the family, she intensified her searches and sold many more of her fossils. She kept the family in this way for as long as it was necessary for her to do so, but as her fame spread, her business of collecting and selling fossils became far too successful to be abandoned. Indeed, she became something of a legend—indirectly, we remember her to this day, for the tongue twister "She sells sea shells on the sea shore" refers to her. The constant exposure to the weather gave Mary severe rheumatism, but there was nothing that could make her change her way of life—or so it seemed to her.

The Dorset coast was also a popular holiday area and many of the summer visitors shared Mary's enthusiasm for fossils. So she opened her own private collection, her "Fossil Depot," to the public, which provided further income. Today many of the fossils Mary Anning collected are stored in the British Museum (Natural History). The Fossil Depot no longer exists, but the building in which it was housed is still there, a few yards from the shore, in the town of Lyme Regis.

In about 1810 Mary discovered in the cliffs the bones of what clearly had been a large marine animal unlike any modern animal. The bones were still articulated and from them it was possible for scientists to attempt a reconstruction of what the animal which possessed them had been like in life. It was later given the name "ichthyosaur."

The ichthyosaur as it was reconstructed was like a fish, in that it was the general shape of a fish, but it possessed limbs rather like those of a marine mammal, and also like a marine mammal, it evidently breathed air, with lungs rather than gills. In other important respects, however, it was clearly a reptile. Thus a picture emerged of a large reptile that looked something like a fish and something like a modern porpoise and that once lived in the sea: a "fish lizard." What was this animal, whence had it come, and where did it live now? Mary Anning's ichthyosaur became justly famous. Mary had found the body. At least, she had found *one* body.

It was not the only curious animal whose remains had been found. It was not even the first. In 1770, in Maastricht, a town in Holland on the Maas (Meuse) River, the jaws of a huge lizard were discovered in a subterranean chalk cave. Later named a "mosasaur" (Meuse lizard), the creature to whom the Maastricht jaws belonged proved to be those of a fiercely carnivorous marine reptile, some 6 to 8 meters long, with a long tail that it used for swimming and limbs modified to form paddles.

In 1784 the remains of a small flying reptile rather like a bat were found, this time in Bavaria. This creature was dubbed a "pterosaur."

Then, in 1822, a Mrs. Mantell discovered some fossil teeth in her garden in Sussex, in southern England. She showed them to her husband, Dr. Gideon Mantell, who was a general practitioner and a keen amateur paleontologist, and to him the teeth appeared to resemble those of iguana lizards he had seen, except that they were very much larger. Then bones were found not far away that seemed to have belonged to the owner of the teeth. Assuming that the remains were those of an iguanalike reptile, Dr. Mantell proceeded to "reconstruct" the owner. His findings, which he published in 1825, described what he called an "*Iguanodon*," a reptile generally similar to a modern iguana lizard, but which was bipedal (two-legged) and stood 10 meters (33 feet) tall. While Dr. Mantell was busy reconstructing his *Iguanodon*, using other paleontological finds to assist him, The Reverend William Buckland, professor of geology at Oxford University, was performing a similar reconstruction of an animal, called "*Megalosaurus*," which was a formidable carnivore. The bones of this animal had been found in Oxfordshire in 1824. Interest in all these discoveries was intense and so was the search that was conducted for other remains. Many were found, and by 1841 nine genera of these lost animals had been identified. Thirty years later, between 1877 and 1880, Dr. Mantell's *Iguanodon* was joined by a further thirty-one iguanodonts, all related, that were found near the town of Bernissart in Belgium.

It was Sir Richard Owen, a British anatomist and zoologist, who devised in 1841 a name for this order of animals. He joined together the Greek words *deinos*, meaning "terrible," and *sauros*, meaning "lizard," to produce the word "*Dinosauria*," and thus the form "dinosaur." The bodies had now been given a name, but had they been identified?

At various times throughout history, people had found what appeared to be gigantic bones. These were explained—because such finds had to be explained—as the remains of giants. It was supposed that at one time a race of human—or we might say "humanoid"—giants walked Earth, accompanied by giant versions of other modern animals. One famous fossil was identified, apparently positively, as "Noah's raven." If the truth were known, it is more than likely that all our legends of giants—and, in the most favored fossiliferous regions, of our own giant ancestors—are derived from the discoveries of such bones, together with, here and there, preserved footprints. The bones and the footprints, then, presented no great philosophical difficulties for people of former times. Nor did they present any greater difficulty for most people in the eighteenth and early nineteenth centuries. It was believed then that all species had been created in a single act of creation and that most had been destroyed in the Great Flood, those which survived being the ones protected by Noah.

They disregarded some of the difficulties Noah must have faced. His ark would have had to find room for something like 1 million species of insects, 25,000 species of birds, 2,500 species of amphibians, 6,000 species of reptiles, and several thousand species of mammals, not to mention cultures of *tens* of thousands of species of microorganisms, all identified without the aid of a microscope. Noah had had to carry two specimens each of all the "unclean" animals, but of the "clean" ones he was commanded to convey seven pairs. All of these had to fit inside an ark whose volume was about 1.5 million cubic feet (43,800 cubic meters), together with their food, of course, which in some cases consisted of live animals of other species, in other cases of particular plants, and in only a few cases, of stored grains.

Be that as it may, the discovery in the late eighteenth and early nineteenth centuries that the bones, footprints, and other remains were not those of giant humans but of quite different animals, led primarily to speculations about conditions on Earth prior to the Flood. In those days the accepted view was that the creation was completed at 9 A.M. on Sunday, October 23, 4004 B.C. That famous date is all that most people remember of the talented and highly educated man who calculated it, Archbishop (of Armagh) James Ussher (1581–1656). He had counted the generations, using the "begats," from the Book of Genesis up to the birth of Christ. Did this make the good archbishop the first person to construct a mathematical model to describe reality? If so, perhaps his modern descendants should be less eager to mock him than they are. In any event, the date 4004 B.C. was well established by the 1800s and it was a brave person who challenged it.

In the end it had to be challenged—by scientists. Science uses models, of course, but it differs from other model-using disciplines in testing its models against phenomena that are observed in the real world, and, at least in theory, by being ready to abandon models when their descriptive power is seen to be inadequate. The first people to question this particular creationist model were the geologists. In the nineteenth century the study of rocks became a popular pursuit for many people, but while amateurs collected data and specimens, the professionals were considering the implications of their finds.

Back in 1785 James Hutton (1726–97), a Scottish doctor turned geologist, presented to the Royal Society of Edinburgh a paper in which he enunciated the theory of "uniformitarianism." This proposed that the geological processes observed taking place on the Earth at the time had operated in a similar fashion throughout the history of the planet. Thus, if we observed a volcanic eruption, for example, and examined the rocks it produced, we would be justified in deducing the existence of former volcanoes from the existence of similar rocks that we observed but in respect of which no recent eruption had been recorded. Hutton assumed, then, that where the process was known by which a rock was produced, that same process had been responsible for the production of similar rocks throughout history. The importance of this theory consists in its rejection of catastrophism or any other singular act of creation or destruction as necessary components of any explanation for the rocks we can see around us today. Later scientists took Hutton's idea further and elaborated on it.

A little while later William Smith (1769–1839), a surveyor who traveled throughout Britain examining landscapes on behalf of companies that were building canals, observed that rock strata always occurred in the same order and that particular fossils could be used to identify layers in sedimentary rocks.

When these and similar observations by other geologists were considered together, certain implications were rather obvious. From the examination of sedimentary rocks it could be deduced that the rocks were formed from accumulations of sediment in which biological organisms were sometimes embedded and that the rocks were then compressed by the weight of rocks above them. Sometimes further processes occurred, and the strata were deformed, or heated, or exposed to liquids or gases that generated chemical changes within them. It could be seen that these rocks had formed during distinct periods when sediment accumulated, punctuated by periods in which no further material was added. At the same time, it could be observed that similar processes still occurred. There were many places where sediments were being deposited and the rate of deposition could be measured. From such measurements, calculations could be made of the time that would be required to produce a layer of rock of a particular thickness, and this calculation then applied to actual rock strata. The result was that the age of rocks was seen to be much older than had been supposed previously. Archbishop Ussher's date had to be abandoned. The age of the Earth was to be measured not in millennia, but in millions, tens of millions, *hundreds* of millions (and as we now know, and can support by more precise radiometric dating methods, thousands of millions) of years. This had implications for the fossil monsters, too, since it was now clear from the rock levels at which their remains were found that they existed a very long time ago indeed. The orthodox view of the natural world had to be, and was, modified.

Then it was observed that the remains of ancient animals bore resemblances to modern animals that were very difficult to explain by a theory of catastrophism and creation. When such observations were joined by rigorous observations of relationships among living species, the hypothesis of evolution by natural selection was developed and, little by little, accepted.

So a cascade of mysteries had been solved—more or less. The old bones had been identified as belonging to a group of extinct reptiles. The age of the bones had been established as tens of millions of years before the present time. The appearance on Earth of the reptiles, and also of the modern species that succeeded them, could be explained by natural selection. A body or two had been found and this had led to further, more thorough, searches being made, as a result of which the one or two bodies proved to have been many bodies. What is more, because the circumstances that lead to the preservation and fossilization of the remains of a plant or animal are so unusual, so special, we can be very certain indeed that the fossils we have found represent no more than the tip—and the unrepresentative tip at that—of a very large iceberg.

With the help of the geologists and the dating methods they devised, it was now possible to date the fossils. The situation at this point was that the bodies had been identified and the time of their death had been established: with one minor exception they disappeared from Earth about 65 million years ago.

Further investigation revealed yet another, even more tantalizing mystery. It could be agreed that the great reptiles were now extinct —although early science fiction writers provided excellent popular

Scelidosaurus

Tarbosaurus

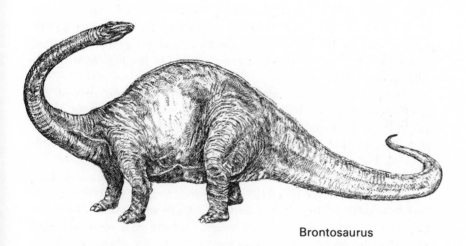

Brontosaurus

Three dinosaurs typical of those which disappeared from the earth 65 million years ago in the Great Extinction. Source: *A New Look at the Dinosaurs* by Alan Chariq. Heinemann/British Museum (Natural History) 1979.

entertainment by supposing them to have survived in certain remote "lost worlds"—but why did they become extinct and, more particularly, why did not only the reptiles but a large proportion of all species become extinct quite abruptly 65 million years ago? Quite simply, the question was: What killed them? And there was no obvious answer.

To those of us who prefer our monsters living rather than dead, the question of their demise may seem less important than the manner of their life. After all, it is far more fun to imagine what these animals may have been like, to try to picture ourselves confronted by *Tyrannosaurus rex* (which in fact disappeared many millions of years before our hominid ancestors appeared) than it is to speculate about its death. Human curiosity is insatiable, especially where something very like a criminal investigation is concerned. We want to know, feel we have a right to know, that justice *requires* that we know, the cause of the dinosaurs' death.

Over the years many eminent men and women have considered the problem and have sought solutions to the mystery. Eventually a point was reached beyond which further speculation seemed fruitless. There were so many possible explanations that there was hardly room for more. The trouble was that there was very little evidence on which the explanations could be based. All that was known with certainty was that the extinctions had occurred. Explanations had to be based on conjecture, and an explanation that seems plausible to one person today is liable to be dismissed by another person tomorrow.

The mass extinctions mark the boundary between two of the great divisions of geological time. The Mesozoic Era began about 225 million years ago, and the reptiles flourished during the whole of it. It is divided into several periods, the last of which is the Cretaceous. The Cretaceous was followed, 65 million years ago, by the Tertiary Period, which is the first division of the Cenozoic Era. Whatever caused the extinctions is known as the "Cretaceous-Tertiary boundary event." The boundary is quite real, something physical. You can see it if you know where to look, as a clear line that separates one layer of rock from the layer of rock above it. (Mary Anning did not find her fossils at the boundary—they came from strata from the Lower Jurassic Epoch, a subdivision of the Jurassic Period preceding the Cretaceous, which made her fossils between 65 million and 200 million years old.) There are places where the boundary can be seen, though. In England, for example, it is exposed in the low cliffs on the seashore at Studland, Dorset, not very far from Lyme Regis but beyond Mary Anning's range.

Although it seemed that way, had the extinctions in fact occurred suddenly? The kind of explanation we need for a sudden event will be quite different from that appropriate to a more gradual event. The difference is similar to that which distinguishes death by violence from death after a long illness. The first question to be answered, then, concerned the speed with which the extinctions took place. The explanations that have been advanced are divided, some supporting gradual extinctions, others supporting sudden extinctions.

Some people supposed that the cause may have been the change in climate that is known to have occurred during the Cretaceous Period. It was believed that an earlier change caused similar mass extinctions at the beginning of the Mesozoic Era and the end of the preceding Permian Period. Those extinctions seem to have occurred suddenly, but if the Cretaceous extinctions were due to such a change, they were more likely to have occurred gradually, because the change itself occurred gradually. In this case the apparently abrupt Cretaceous-Tertiary extinctions may be no more than the misleading product of sparse fossil evidence. The record available to us may refer only to the species that disappeared over a short period and those which disappeared over a longer period before that may not have been preserved or their fossil remains may not have been found.

More recently it has been discovered that from time to time Earth's magnetic field reverses its polarity, north becoming south and south north. While the reversal is taking place—and it happens over a period of a few thousand years—the magnetosphere, the shield formed by the magnetic field that envelops the planet, is weakened. Such weakening might permit the entry of ionizing radiation from the Sun. It used to be believed that this would reduce the layer of ozone that traps much of the incoming ultraviolet (UV) radiation and so irradiate the surface. Are we really to suppose that the species that died out were fried in UV? It has been suggested that polarity reversals may be caused by meteoric impacts. In this case, rather than being the cause of a biologically important event, is the polarity reversal merely an indicator of it?

A more mundane explanation of the extinctions involved the rise of the mammals. Being "warm-blooded" animals whose body temperatures were maintained independent of the outside environment (homoiothermic), the mammals could pursue a nocturnal way of life, whereas the "cold-blooded" reptiles, whose body temperatures were dependent on the outside environment (poikilothermic), could not. If the ancient reptiles were like most of their modern counterparts, it is possible that their unguarded eggs fell prey to mammals that fed at night, which so reduced the reptiles' breeding efficiency and depleted their numbers that they died out completely.

Or perhaps a star exploded as a supernova in our region of the galaxy, showering the surface of the Earth with high-energy particles. Did species die from such radiation, directly or indirectly through being rendered infertile? Is that what happened? Did they therefore die rather abruptly?

Maybe a cloud of interstellar dust moved through the solar system (or the solar system moved through the dust, which amounts to the same thing). Did this blanket out solar radiation just for a short time, making the world dark? In a dark world, green plants, which depend on sunlight, would die and animals that feed on them would starve. In any case, a dark, temporarily cold world would be an inhospitable place for poikilotherms. Did the Late Cretaceous sky grow dark?

Then there is a final possibility, and the most dramatic of all. Did some large object—a meteorite, say, or a comet—crash into the Earth? If so, what effect would that have had? "Planetesimal" is the general term used for small solid objects that exist in space and are the size of a tiny planet—a dozen or so miles in diameter. Meteorites and asteroids are planetesimals and travel at high speeds. Comets are smaller and composed half of ice, but they travel a great deal faster and so may possess more or less the same energy.

If a planetesimal had entered the Earth's atmosphere at a speed of something like 20 kilometers per second (about 45,000 miles per hour), it would have taken only a few seconds to reach the surface from outer space. The impact would probably have sounded like a huge explosion. To compare such an explosion to that of the largest existing H bomb might be like comparing the eruption of Mount St. Helens with the firing of a child's cap gun. It would have been far beyond anything human beings have ever experienced. The sound would be carried as one of several sets of shock waves and the earthquakes caused by the impact would have been registered around the world. If the object had impacted into the sea, there would have been tsunamis (so-called tidal waves) racing across the oceans at hundreds of kilometers an hour, building up into great walls of water perhaps hundreds of meters high, as they swept across the coasts of continents. An impact on land would have formed a crater perhaps 200 kilometers in diameter, and material would be ejected into the atmosphere, perhaps even into space, from where it would reenter the atmosphere gradually, over a long period, to fall as dust on Earth. The passage through the atmosphere of a body possessing so much energy would have caused chemical reactions in the air itself; more chemical reactions might follow the release of dust and possibly magma—fluid rock below the Earth's crust—vented from volcanoes that erupted because of the violent shaking they would have received.

On the other hand, if this object had entered not at a steep angle and fast, but at a shallow angle so that it ricocheted off the surface of the Earth and was captured in an unstable orbit, at each pass some of its energy would have been lost so that it would have moved ever closer to the Earth until at last, broken by now into fragments, it fell to Earth and was scattered over a wide area. In the course of all this its vast energy would have been expended, albeit more slowly, in the Earth's atmosphere.

Would such a collision have killed off whole groups of species? It might have. A blanket of dust that shaded the Earth's surface from sunlight could cause many deaths, as we suggested in the case of the cloud of interstellar dust. Chemical reactions could produce similar darkening if large amounts of nitrogen oxides were formed; if the products of these reactions included nitric, hydrochloric, and sulfuric acids, as they may well have done, their precipitation in "acid rain" would cause great harm to vegetation.

In fact there are several ways that an event such as this might account for the extinctions, although some of the possible explanations—such as starvation caused by the death of food plants—may compel us to consider the possibility that the last of the great reptiles included many species that were homoiothermic.

This is all very well, but is there any reason why we should prefer this catastrophic explanation of the extinctions to those that have gone before? Is it based on anything more than guesswork?

To answer that we must follow the trail of clues by which scientists have moved in the last few years toward the very tentative conclusion that the Cretaceous-Tertiary boundary event was caused by a collision between the Earth and an extraterrestrial body.

2

Clues from the Clays

There were scientists scrabbling about in the earth. They were not seeking anything. They were taking samples that would be analyzed in laboratories. They hoped that these samples and their chemical contents would provide them with a better method than had existed previously for dating particular rock strata.

They were working near the town of Gubbio, in central Italy. That is its modern name. The people of ancient Umbria called it Iguvium, and so did the Romans who built its theater. Later, in the fourteenth century, it acquired a splendid palace, the Palazzo dei Consoli, which today serves as the town hall.

Gubbio is nestled in the Apennine mountains, in the province of Perusia. It is a charming, pretty town, the home of some 32,000 people. Like some other Italian towns, Gubbio is proud of the medieval traditions it preserves. For centuries it was independent. It was not until 1860 that it was united with the kingdom of Italy. The town dominates a large plateau at the foot of Mount Ingino, high above sea level, where the climate is temperate. The fields around Gubbio produce grapes from which a local wine is made, olives for oil, and cereals.

As early as the sixteenth century, though, Gubbio was famed throughout Italy for its ceramics; more recently cement and brick have become important commercial products. The ceramics and the brick are made from local clays, and it is these clays that interested the scientists.

They were interested because the Gubbio clays mark, very clearly, the boundary between the Cretaceous and the Tertiary periods. Gubbio is one of those places in the world where the boundary can be seen plainly. If the chemical composition of the clays was in some way peculiar to that stratum, then the discovery of clays of a similar composition elsewhere in the world might be taken to indicate the same boundary. The clay would stand as a marker by which people could identify positively a point in the Earth's crust 65 million years old and so below that marker the rocks—and any fossils they contained—would be older than 65 million years and above it they would be younger.

Perhaps the word "clay" requires qualification. To a geologist it signifies nothing more than a soil composed of particles that are less than 0.004 millimeter in diameter—about half the width of a red blood corpuscle. If clay marks the bottom end of the scale of soil particle sizes, at the opposite end there is sand, whose particles are from 0.05 (fine sand) to 2 millimeters in diameter. So clay and sand tell us nothing whatever—well, nothing very much—about the *chemistry* of the materials from which they are made. To a geologist, any mineral earth material is "rock," so there is nothing inconsistent about using the word "rock" interchangeably with "clay" or "sand."

The scientists were looking at these particular Gubbio clays, and they found that the clays contained relatively large amounts of iridium and osmium.

Iridium and osmium are precious metals, precious because they are very rare. They belong to the group of metals called "platinoid" because the most famous member of the group is platinum. Apart from their relationship within the platinoid group, iridium and osmium are also known as "noble metals," because of their reluctance to enter into chemical reactions and so to become "defiled." Noble metals survive the acid test, which means they are not attacked by simple acids but that they can be dissolved—in the case of iridium and osmium not readily—by aqua regia, a mixture of nitric and hydrochloric acids. The other noble metals include gold, platinum, and palladium. Because they are noble, they are often found (except that they are not often found at all) as pure metal.

The scientists at Gubbio were interested mainly in iridium and osmium. The other noble metals came to excite them later.

Iridium is a white metal and it is very heavy—in fact, iridium and osmium are the densest of all terrestrial substances. Iridium was discovered in 1803 by the English chemist Smithson Tennant, who named it from the Greek word for "rainbow." It has a specific gravity of 22.36 grams per cubic centimeter at 20 degrees C. Its atomic number is 77 and its atomic weight 192.22. It melts at more than

2,400 degrees C and boils at more than 4,100 degrees C. It is used, as an alloy with platinum, in jewelry, in some electric points where its resistance to the high temperatures generated by electric sparks is useful, in surgical pins and pivots, and in the nibs of fountain pens. It also supplies part of the metal—10 percent alloyed with 90 percent platinum—for the international prototype meter and kilogram.

Osmium occurs in the mineral osmiridium and it is found in most platinum ores. Its name was also given to it by Tennant in 1803, who separated it for the first time. The name comes from the Greek word meaning "smell," because one of its compounds has a bad smell, not unlike chlorine, and is no less poisonous. As a metal osmium is hard, brittle, silver-gray in color, and almost impossible to work. Its specific gravity is 22.59, its atomic number 76, its atomic weight 190.2. It melts at a little over 3,000 degrees C and boils at 5,300 degrees C. Like iridium, sometimes it is used to harden pen nibs. More useful, perhaps in this age of the felt-tip, it is used to stain lipids (fats) in biological materials to be studied under the optical or electron microscope.

These metals, then, are very dense, very heavy, and very rare at the surface of the Earth. There are places in the solar system where they are less rare than they are on the surface of our planet, though they are never common.

The fundamental building materials of the universe are hydrogen, the lightest element, and helium, the second lightest. Elements heavier than these can be formed only under the conditions that obtain inside stars, where thermonuclear furnaces provide great heat. They are, in the most literal sense, the by-products of fusion reactors. Hydrogen nuclei are fused to form helium, then hydrogen and helium nuclei are fused to form heavier nuclei and these in turn are fused to form the nuclei of elements that are yet heavier. The process continues until iron is formed, but there it ends in stars like the Sun. It is only in the hottest of all stellar furnaces that elements heavier than iron can form. To make iridium and osmium requires temperatures that exceed those achieved during the main part of the lifetime of a typical star. It requires massive stars, which have exhausted their stock of hydrogen and helium and which begin to burn such "forbidden" elements as carbon and neon. This initiates fusion reactions that are inherently unstable and, provided the mass of the star exceeds a certain minimum value, the star explodes to become one of the two possible types of supernovae. Iridium, osmium, and all the very heavy elements were manufactured just before and during a brief, spectacular, and extremely hot supernova event.

After a star has exploded in this way, typically its core contracts and it enters its old age. Depending on its mass this may form a neutron star or even a black hole. The outer material that was cast off during the explosion does not return to it, however. It moves away rapidly into space, as a cloud of hot particles of various sizes. There are also, in space, clouds of dust and particles that are cool. The Crab Nebula is such a cloud. Should a hot, supernova cloud interact with a cool cloud that is denser than itself, the shock waves generated by such an interaction could cause both clouds to condense and form a new star, and that star could possess planets, formed partly from the raw material supplied by the supernova explosion.

The solar system is composed, in part, from such a mixed cloud. Many of the atoms on Earth originated in that primordial thermonuclear holocaust. Since all flesh comprises these same atoms, we are made partly from fallout!

As the Sun and its planets condensed, the particles of the cloud being drawn together by gravitational attraction, a stage was reached at which the elements from which the system was composed were distributed more or less evenly. As the system condensed further, the elements within it became sorted. The lighter elements moved toward the outer planets. The inner planets were heated strongly, the heating causing most of the lighter elements to be lost into space while the heavier elements began to coalesce around cores formed from the heavier metals, which were molten. The central core of the Earth is composed mainly of molten iron and nickel, and the other heavy metals, including iridium and osmium, which have an affinity for iron, were absorbed into the core as the solid planet formed.

If that seems reasonable, the question that follows naturally is why did some of the iridium and osmium remain at the surface? It is believed that where we find metals like iridium and osmium at the surface—or, today, near the surface in rocks that once lay at the surface—they have arrived from space. But why do they appear to be more abundant in other regions of the solar system than they are on Earth? In fact they are not. When the original cloud condensed, not all of it was used in the formation of what are now the Sun and planets. Some dust remained, which now comprises the asteroid belt. Most astronomers believe that the gravitational tides exerted by the immense mass of Jupiter prevented a planet from forming from this dust. There probably was insufficient material available to form a planet of a reasonable size, but the debris that failed to consolidate itself exists now either as dust, some of it dispersed through the system, or as the asteroids themselves.

The asteroids are believed to be the most likely source of meteorites—although there are scientists who disagree—and meteorites may be divided into two principal types. Those that are made mainly from rock—"stony" meteorites—are believed to consist of the primordial material from which the solar system is made. The metallic meteorites—"iron" meteorites—may have been heated strongly at some point in their history, or they may represent the cores of objects that once were much larger but from which the outer, stony layers have been lost, probably by impacts with other meteorites. These iron, or siderophile, meteorites usually contain noble metals at concentrations much higher than those found either in stony meteorites or in surface rocks on Earth.

Because the elements were distributed more or less evenly throughout the original cloud, the abundance of any particular element will be the same at any point, at least in the inner part of the solar system, but because heavy elements are locked in the planetary cores, the crustal rocks appear to be depleted. The cloud material that never formed a star or a planet has suffered no such apparent depletion, either because it has no core or because it consists of nothing but a core. Therefore matter that reaches the Earth from space is likely to have a composition much like that of the original cloud. Deposited on the Earth's surface, it is able to supply the rare elements that are trapped in our planet's core. This background information is necessary for an understanding of the importance of the clue the scientific detectives found at Gubbio.

The team of detectives at Gubbio was led by Walter Alvarez, professor of geology at the University of California at Berkeley, and it included Alvarez's father, Luis W. Alvarez of the Lawrence Radiation Laboratory at Berkeley and two nuclear chemists from the same laboratory, Frank Asaro and Helen V. Michel. Luis Alvarez won the 1968 Nobel Prize for physics for his work with elementary particles which included the discovery of resonance states. It was, in short, a high-powered team.

They had a fairly new analytical technique to help them: neutron activation analysis, or NAA. Just how this works is less important to the story than the fact that it makes possible the accurate measurement of concentrations of elements that are present in very small amounts indeed. This is more of a problem than it may at first seem, because of what geochemists mean when they talk of an element being "rare."

In average granite, for example, you may expect to find 0.0001 part of osmium, and 0.006 part of iridium for every million parts of rock. In other words, to concentrate 1 gram of osmium you would need to process 10,000 tons of rock, and you would need to search through 6,000 tons of rock to find one gram of iridium. Admittedly, granite is not an especially good place to go looking for these metals, for it contains less of them than some other rocks, but even so, the average crustal concentrations of osmium and iridium are about 0.005 part per million and 0.001 part per million, respectively. Until the NAA technique became available, the accurate measurement of concentrations as low as these was notoriously difficult and uncertain. In the solar system as a whole, on the other hand, both elements are about a thousand times more abundant and where samples of the original solar system raw material can be found—from dust or from meteorites, for example—concentrations may be much higher than they are on the actual surface of the Earth. The Gubbio team's analyses began with iridium because even with the NAA technique, iridium is easier to identify and measure than most other members of the platinoid group.

The first samples were taken from the Gubbio clays, and sent back to the Lawrence Radiation Laboratory in Berkeley for analysis. The analysis showed that the clays contained about 30 times more iridium than the strata immediately below or above the 65-million-year level.

This raised a question. Was the enrichment of the Gubbio clays a purely local event or might such an enrichment be found at the same level in other places? The scientists needed samples of boundary clays from other sites and they went first to a place about 50 kilometers south of Copenhagen and then to New Zealand's North Island, right on the other side of the world.

The Danish samples were taken from a site near Højerup Church, at Stevns Klint, where there are well-known "fish clays"— so called because fossilized fish remains are found in them—formed most probably in a closed basin containing stagnant water. The conditions under which the clays formed are deduced from their chemical composition. Boundary clays elsewhere have a similar composition and it is difficult to imagine how those that formed beneath deep oceans acquired it. Analysis of the fish clays showed that iridium was present in them in concentrations about 160 times greater than the Earth's crustal average. Then the clay samples arrived from New Zealand. They had been collected from a site near Woodside Creek, about 40 kilometers northeast of Wellington, and they were found to contain about 20 times more iridium than the rocks above and below them. Although they differed from the first, Gubbio, samples, the Danish and New Zealand samples confirmed that the iridium enrichment was not a local Italian phenomenon. Clays at the 65-million-year level contained anomalously large amounts of iridium in other parts of the world.

Having determined the concentrations of iridium in the boundary clays, the laboratory analyses went on to examine the concentrations of other elements: twenty-seven of them in the Italian samples and thirty-three in the New Zealand ones. They found it was not only iridium that existed in anomalous amounts. So did most of the other metals. It was as though the chemistry of those clays—from widely separated locations—had been skewed.

Walter Alvarez and his team published their results first in 1979 in the *Geological Society of America: Abstracts with Program* (Vol. 11) and then in 1980 in the American journal *Science* (Vol. 208, pp. 1095–1108). At once there was the flurry of interest in the scientific community that is created whenever a new, plausible, and possibly exciting idea emerges. People began to check the Alvarez findings and as they did so they found themselves merely adding more supportive analyses from many new sites.

The existence of a clear layer of these enriched clays marking the boundary between the Cretaceous and Tertiary was well established. It had been identified positively in Zumaya, in northern Spain; at El Kef, in northern Tunisia; at many oceanic sites where cores of material had been obtained from deep beneath the sea floor in the Pacific by workers on the United States research ship *Glomar Challenger* in the course of the international Deep Sea

Drilling Project (Leg 62, Hole 465A, but other Pacific cores were examined by others); and, in the greatest possible detail, at Caravaca in southern Spain. The fact that the boundary can be seen so clearly in so many different places is an important item of evidence, and we shall return to it later, in Chapter 6.

R. Ganapathy, of the research laboratory of the J. T. Baker Chemical Company of Phillipsburg, New Jersey, visited the fish clays in Denmark, taking two samples and analyzing their contents for the noble metals and comparing his figures with those obtained from analysis of seabed clays and basalt obtained from the Columbia River. His results, published in *Science* in August 1980 (Vol. 209, pp. 921–23), showed that "all these elements are present in the boundary clay at levels well above those normally seen in such terrestrial materials as continental basalts, oceanic basalts, and pelagic clay." In fact, he found up to 69 parts of iridium per billion parts of other matter; iridium, osmium, and palladium were all in concentrations more than 3 times richer than normal.

J. Smit, of Amsterdam, and J. Hertogen, of Louvain, Belgium, went back to the Caravaca clays to examine them in greater chemical detail. They took a hundred samples and considered a total of twenty-seven elements. The samples were taken from different levels, so Smit and Hertogen were able to relate their findings to a sequence in the clays themselves. What they discovered was that all the elements occurred in the concentrations that would be expected through most of the layer they examined—they were present in their typical terrestrial abundances. Then, in the last few centimeters, they found the amount of iridium was 450 times the crustal average, osmium 250 times, and arsenic 110 times. Chromium was 9 times richer in the clays at the same level, cobalt 30 times richer, nickel 44 times richer, selenium 40 times richer, and tin 20 times richer. It was at precisely this level, they said, that there was a great depletion in the number of fossils. Smit and Hertogen published their findings in the British journal *Nature* (Vol. 285, pp. 198–200, 1980).

Frank T. Kyte, Zhiming Zhou, and John T. Wasson, all from the Institute of Geophysics and Planetary Physics of the University of California at Los Angeles, also published similar findings in *Nature* later in 1980 (Vol. 288, pp. 651–56). Like Smit and Hertogen, they had been excited by the early Alvarez reports and set out to examine clays for themselves, this time in two places. They took twenty-three samples from Denmark, using the Stevns Klint fish clays again, and three from one of the deep-sea cores in the Pacific— from Hole 465A, in the center of the ocean, roughly halfway between Yokohama and San Francisco.

Once again the results confirmed those from other sites. Although the deep-sea core samples contained much less iridium than samples taken from on land, they showed a sudden sharp increase, from less than 0.7 part per billion at one level to 12.9 parts at another and then to 15.6 parts in still another.

We can now compare the concentrations of both iridium (Ir) and osmium (Os), expressed in parts per billion, in normal Upper Cretaceous rock, in the Stevns Klint clays (Kyte team), and in those at Caravaca (Smit and Hertogen team):

	Upper Cretaceous	Stevns Klint	Caravaca
Ir	0.13	35.7	25.5
Os	0.08	43.8	16.1

The concentrations of many other elements showed similar contrasts.

Immediately it is also clear from the above table that the Danish clays contain more iridium than those found at other sites. This presents another anomaly that seems to require explanation. Or does it? Imagine first that dust is falling onto the surface of still water. It will drift down through the water to form a part of the bottom sediment, where it will comprise a certain proportion of that sediment. Consider now the case of dust falling onto more open water where currents transport material. In this case the sediment and its load of dust will be moved from one place to another so that in some places the concentration of dust will be higher than in others. Might this explain the apparent anomaly between concentrations of metals found in the Danish fish clays and those found elsewhere? It is a question we will explore in more detail in a later chapter.

The Alvarez team began their original investigation because it was suspected that the clays marking the Cretaceous-Tertiary boundary could be indentified chemically. They and other workers who followed them went on to discover not only that this is true, but that the boundary can be identified clearly in many different parts of the world in both hemispheres. This is unusual. The Earth is constantly restless. Rock strata are continually deformed, twisted, heated, and cooled, and the constituents of most strata become dissipated in surrounding materials. Sedimentary material, sinking to the sea bed, may be stirred and mixed by any one of a number of mechanisms so that it merges with the sediment below it, and layers grade into one another rather than show sharp divisions. Sharp divisions occur when sedimentation ceases for a time and the material is compressed before it resumes again so that mixing between what is now soft rock and sediment is impossible. In the boundary clays, however, the concentration of noble metals is confined very precisely. The enrichment occurs in a band whose width can be determined, so that the composition of the material a centimeter to one side of it reflects the average terrestrial abundance of elements, but within it these amounts change dramatically. You cannot see the chemical line in the gray-colored clay with the naked eye, but it is there.

There are natural, terrestrial processes by which particular elements can become concentrated. The most usual process, and perhaps the principal source of many metallic ores, is partly biological. Soluble metallic compounds travel through ground water until they reach the bottom of pools of stagnant water. The sediment below this water consists of muds in which organic detritus (the accumulated wastes and remains of once-living organisms) is undergoing anaerobic bacterial decomposition, as a by-product of which such gases as methane and hydrogen sulfide are produced. The action of these gases on the muds tends to separate out insoluble metallic compounds, which are precipitated and concentrate in the sediment, later to become ores.

Some processes are entirely biological, when microorganisms extract nutrients from a substrate containing metallic compounds and concentrate the metals in this way. It is also possible for a metallic ore to be harder than the rock surrounding it. If such rock lies close to the surface, ordinary weathering may remove it, leaving the ore behind. If this happens, a rock sample taken from the surface may well show a high concentration of a metal that otherwise is rare—but it has been concentrated naturally. Not only is this possible, it has happened and noble metals are known to become concentrated in this way. Ganapathy, who had analyzed the Danish clays and clays from the Columbia River, drew attention to Canadian mines where ruthenium, osmium, iridium, palladium, and gold are obtained from ores that are wholly terrestrial in origin. These metals sometimes occur in copper-nickel ores and in molybdenum sulfide ore, their atomic configuration allowing them to lodge in the crystal lattice of the rock. Where this happens, the enrichment is local, and the biological processes are also local, of course.

In the case of the boundary clays, the enrichment is unlikely to have been a local phenomenon—it would be too much of a coincidence to find the same local terrestrial enrichment process taking place more or less simultaneously in so many different places. Furthermore, when noble metals are concentrated by terrestrial processes other than the biological ones, it must be because they are contained in the crystal lattices of particular rocks which have particular chemical compositions. Thus the noble metals are inevitably accompanied by other elements. The presence of these associated elements can be predicted, and so can the proportions in which they will be found. In the boundary clays the proportions are quite wrong.

We must conclude, then, that the high concentration of iridium in the boundary layer is not due to any terrestrial process of which we have knowledge. The Alvarez team suggested that the iridium must have come from space since its worldwide distribution shows it is most improbable that it could be due to local enrichment. They pointed out that about once every 100 million years we should expect the arrival of an object from space (that is to say, from somewhere beyond the Earth's atmosphere) that would provide the requisite quantity of noble metals. Smit and Hertogen took a similar view, arguing for an extraterrestrial source because terrestrial sources seemed inadequate to account for the high concentrations of iridium and osmium. Ganapathy was impressed by the proportions of the elements found in the clays he examined, which provide a further reason for supposing them to contain extraterrestrial material. The Kyte team was more cautious, but in the end they too favored an extraterrestrial source.

If the metals came from space, in what form did they arrive? There are four possible alternatives: a dust cloud, a cloud of matter caused by a supernova explosion, a comet, or a meteorite. Much depends on whether the material arrived in a fine, dispersed form so that it settled as dust might settle or arrived as one or more large masses.

So far as Ganapathy was concerned, the osmium could be used to test the idea of a supernova origin. Osmium occurs in two isotopic forms relevant to our story, osmium 184 and osmium 190. (The atoms of many elements exist in several forms, each with a different atomic weight. Because the chemical properties of these atoms are very similar, clearly they are atoms of the same element, but nevertheless they are slightly different. Each of these "species" of an element is called an "isotope" of that element. The distribution of isotopes of any particular element is constant, so that tin, for example, consists of ten isotopes, always the same ten and always in the same proportions.) According to Ganapathy, the proportion of one osmium isotope to another is constant in terrestrial samples and in samples found in meteorites. In different supernovae, however—and remember that all osmium is formed in supernovae—he claimed that the proportion of one isotope to the other of the same element may be expected to vary. In his view, we find a particular proportion in the solar system because the matter from which the entire solar system is constructed originated in a single supernova. If the osmium in the clays came from a different supernova, he said, it would be a strange coincidence if they repeated the proportions that were formed in the earlier explosion. Ganapathy measured the two osmium isotopes in his Danish and Columbia River clays and found the isotopes to be present in the ratio typical of terrestrial and meteoric osmium. That would seem to rule out a different supernova and thus a source outside the solar system. The metals, he concluded, came from somewhere *inside* the solar system.

We should emphasize that this view of the reason for the distribution of isotopes is the opinion of Dr. Ganapathy. He may be correct, but as we shall suggest later, an alternative view is no less tenable.

An interstellar dust cloud could contain the requisite amounts of iridium and osmium, and about once in every 100 million years the solar system probably does encounter such a cloud as it passes through an arm of the galaxy. The amount of material that would enter Earth's atmosphere and settle on the surface would depend on the density of the cloud, the speed at which it was encountered, and the length of time during which the Earth passed through it— which in turn depends partly on the size of the cloud and partly on the direction in which it moved in relation to Earth's orbit. Kyte and his colleagues calculated the size of cloud that might be required to deliver the amount of material that seems to exist. This turned out to be a cloud with a density of 100,000 hydrogen atoms per cubic centimeter (which does not mean the cloud was composed wholly of hydrogen, of course—this is merely a convenient way of measuring density) and with a diameter of about 860,000 billion kilometers (5,700,000 astronomical units or 28 parsecs). This is about ten times larger than the largest and densest clouds in our neighborhood of the galaxy. What is more, Kyte calculated that because the iridium is concentrated in such a narrow band in the Caravaca clays, it must either have been obtained from a part of the cloud 10 times more dense than the figure he had allowed or the rate of sedimentation in the sea at that time must have been less than the lowest deposition rate observed today. Otherwise the iridium would have been distributed in a large amount of sediment and the band would have been thicker. So there are great difficulties in supposing that the iridium and osmium came to us from an interstellar cloud.

We are left with the comet and the meteorite. Kenneth J. Hsü, of the Geological Institute of the Swiss Federal Institute of Technology at Zurich, favored a comet (*Nature,* Vol. 285, pp. 201–3, 1980). The event would have been similar to the explosion that occurred in the Tunguska Basin in Siberia on June 30, 1908. That may have been caused by a small comet that exploded and disintegrated in the air. Instead of forming a single large crater it formed many smaller craters (50 to 200 inches in diameter). Hsü suggests that the comet that entered the atmosphere 65 million years ago was about the size of Halley's comet, with a mass of about 1,000 billion tons. It entered with such energy, he says, that it heated the atmosphere and then entered the sea, heating that also. Not much is known of the chemical composition of comets, but Hsü proposed a chemical as well as a physical effect. The comet might have added large amounts of cyanides (hydrogen and methyl cyanide have been detected in the tail of Comet Kohoutek), which would have poisoned many organisms. However, as Kyte pointed out later, it is very doubtful whether cyanides could have survived the passage through the atmosphere where, at the temperatures that must have been generated, they would have been oxidized and made harmless. Even if they had entered the seas, cyanides would have broken down rapidly, would not have dispersed far from the points at which they entered, and despite the stories invented by authors of crime fiction, cyanide compounds are less poisonous than is often supposed. Compared with many common agricultural chemicals, for example, quite large doses are needed to kill most organisms. It is doubtful, therefore, whether cyanides could have been present in sufficiently high concentrations, or for long enough, to cause serious harm.

Finally we must consider the possibility that the object was a meteorite. For it to have supplied the required amount of boundary material, the body must have been 10 or 11 kilometers in diameter, weighing about 2,500 billion tons. Perhaps some of its rivals have not been disposed of entirely, but a meteorite emerges as the most probable culprit, and we will examine it in greater detail in Chapter 5.

Some additional evidence appeared later which showed that whatever the event may have been that deposited the metals, it was violent and generated much heat. Smit and another colleague, G. Klaver, of the same institute (Geological Institute, Amsterdam), returned yet again to the Caravaca clays in Spain and at the base of them they found many very small spherules of glassy material similar to the tektites that are produced when rocks are melted. Such spherules are sometimes found in the vicinity of volcanoes, but chemical analysis showed these particular ones to have a composition inconsistent with a volcanic origin. They suggested (*Nature*, Vol. 292, p. 47, 1981) that the spherules were formed by an impacting body, which may have been an iron meteorite or, less probably, a comet.

The purpose of the original Alvarez investigation at Gubbio was to discover a chemical "signature" by which boundary clays could be identified simply and reliably. The hope was that a convenient means would be found to locate the boundary in rock strata and at the same time to date more accurately rock strata in the vicinity of the boundary. That the chemical signature at Gubbio might have been caused by some extraterrestrial event was interesting, but not especially relevant to the identification. It was only later, when it was realized that the boundary appeared to be marked by an event of huge proportion that scientists began to suspect that the event might not merely coincide with the Cretaceous-Tertiary extinctions and so mark them, but might be the cause of them.

Of course, the fact that large amounts of iridium and osmium arrived from an extraterrestrial source at just the time that so many plants and animals became extinct may be nothing more than coincidence. The evidence thus far is circumstantial. But so is a smoking gun.

The verdict of the scientific jury, returned late in 1980, was that these species had died suddenly under circumstances that suggested an extraterrestrial cause.

3

Gradual or Sudden?

There are two sides to any argument, they say, but the truth is that there are two sides only to those arguments that consist of a single proposition and its opposition. The proposition must be attacked and defended. There is a defense and a prosecution, as there is in a judicial trial. Our argument has not yet reached so simple a stage. There are several alternatives, each of which demands a consideration.

Despite the "verdict" that was reached at the "inquest," the investigation must continue. Perhaps the species of the Upper Cretaceous became extinct in the ordinary course of events. Species do become extinct, after all. How certain are we that their demise was abrupt?

There are two possible mechanisms by which the species in question may have died out gradually. One is competition from already existing species that were better adapted to the environment. The other is change within the environment—a major change in climate or a chemical or microbiological change; they are not mutually exclusive—they could act together—and the picture is complex.

It is a fallacy to suppose that evolution proceeds only by the adaptation of species to preexisting environmental conditions over which they have no control and that environmental change is wrought by physical or chemical forces that act independently of living organisms. There are such forces, of course. But for any organism the environment includes other living organisms, each of which modifies its own immediate environment to a greater or lesser extent. It is fashionable today to consider human behavior as aberrant because it alters the face of the Earth, causes gross changes in "natural" ecosystems, and disturbs the physics and chemistry of the planet. If our behavior is unusual, it is only because of the scale on which it operates. If we imagine a world in which, say, beavers were as numerous as humans are now, clearly they would change the face of their world dramatically. So would almost any species. Elephants push down trees whose edible foliage is beyond the reach of their trunks. They can and do clear forests in this way and thus contribute to the development of grassland. Prairie dogs clear away shrubs from an area surrounding the entrances to their burrows in order to deprive predators of cover and incidentally promote the growth of their preferred food plants.

These are among the more spectacular examples of environmental modification, but since the growth of one plant precludes the growth of another plant on the same spot—which is what we mean by "success" in an ecological sense—and since an animal cannot feed without making at least a minor change in its surroundings, the final list of species that alter the environment must be as long as the full list of species itself. Therefore adaptation to an environment means changing the conditions established by living organisms that have gone before just as much as it means adaptation to inanimate forces. There is good reason to suppose that the entire face of our planet, from the depths of the oceans to the uppermost limits of the atmosphere, has been modified by life itself to an extent so large that we may be justified in considering the entire biosphere as an artifact, formed by microorganisms and, to a much lesser degree, by multicelled plants and animals.

Climate change may be caused by biological activity and it is fairly certain that it has been caused in this way in the past. Biological or not, however, such change could have led to conditions in which certain species found themselves at an advantage. It is no less possible that changes among species themselves—perhaps in micropopulations that have left no fossil records of themselves and of whose activities we can know little, and none of that by direct observation—created altered conditions that some of the larger species found tolerable but others could not endure.

It is a complication in the Upper Cretaceous case that few of the possible explanations for the extinctions necessarily exclude all others. In the end it may prove that while one cause was paramount, others operated at the same time, tending to produce a similar conclusion, and that the waters may never be made crystal clear.

Let us consider first the possibility that competition from other species brought about the extinction of those Upper Cretaceous species which disappeared. It has been said, for example, that the dinosaurs had evolved to a very large size, that they were unwieldy, slow, and stupid, ill equipped to meet any challenge that could not be countered by their own brute force. Had any new group of animals emerged that were less specialized, they might have found it not too difficult to exploit the environment more efficiently than, and so to the cost of, the dinosaurs. Were there such groups? Yes, there were two: the birds and the mammals. If we suppose, then, that the great reptiles, the birds, and the mammals comprised three teams of rivals, it may be useful for us to consider the relative strength of those teams during the latter part of the Cretaceous.

The Cretaceous is one of the longest periods in the geological history of the Earth. We must treat this statement with some caution, however, since the episodes of geological time are inventions of paleontologists and geologists. As such they are convenient, but somewhat arbitrary. The Cretaceous Period, which was named (from the Latin word meaning "chalky") by the Belgian geologist J. B. J. Omalius d'Halloy, comprises strata of sedimentary rocks. The extinctions we are considering mark the period's end. In this sense it could be described as one of the longest periods that we perceive distinctly in the history of our planet.

The Precambrian Era, which occupies about five sixths of the entire history of the Earth, is much longer. This is the era during which life first appeared on the planet and microorganisms modified the seas, the atmosphere, and at least some of the dry land surface in ways that were exploited later by other forms of life. Very little is known about these microorganisms. They left few fossils, and because our dating of Earth history has relied so heavily in the past on the particular kinds of fossils that are found in particular rock strata, it used to be assumed that the absence of fossilized remains of multicelled species in Precambrian rock implied an absence of life then. More recently it has been recognized that not only is this picture untrue—for microfossils have been found—but that it cannot make sense. Microorganisms preceded the appearance of large organisms, and for most of Earth's history they had the planet to themselves.

During their long, exclusive occupation they brought about huge changes—in the chemistry of the seas but most dramatically in the

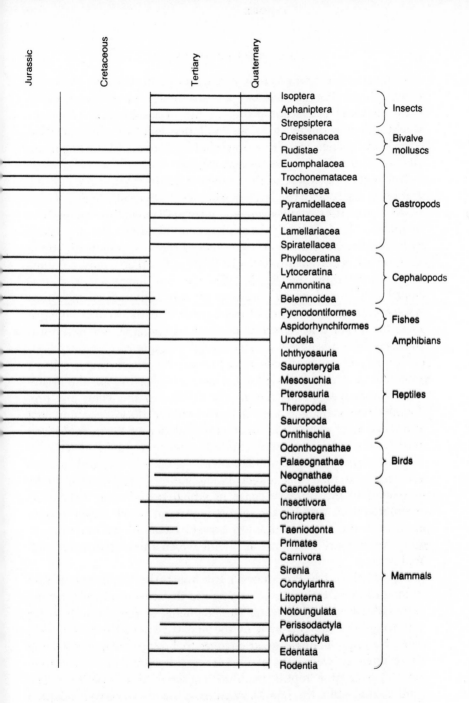

This table gives some indication of the devastation of species in the Great Extinction. Notice how many species disappear at the end of the Cretaceous period, and how many new species spring up. Source: Raup & Stanley, *Principles of Palaeontology.*

atmosphere. For it was as a by-product of photosynthesis by single-celled plants that gaseous oxygen was produced in quantities so great that the gas accumulated in the atmosphere. Beyond any doubt, this was the most serious "pollution incident" the world has ever known, since oxygen is powerfully reactive and, to organisms that are unused to it, extremely poisonous.

Nor is this the whole of the oxygen story, for should the concentration of gaseous oxygen at any time exceed certain limits, its oxidative properties would overcome what resistance the species have acquired. The oxidation of carbonaceous material would proceed very rapidly. In other words, any one of the 10,000 or so lightning flashes that occur each day might ignite a fire that would consume every living thing on Earth. This has not happened and (we hope) will not happen, because of the activities of countless microorganisms that either utilize oxygen directly, so scavenging it from the air, or that release gases, such as methane, which are oxidized in the atmosphere, so "locking up" oxygen.

The role of microorganisms in forming the world that we know must not be overlooked simply because it is obscure, nor the organisms themselves simply because they are small. They are profoundly important and very pertinent to our story. Anything that produced large changes in the micropopulation of Earth could lead indirectly to major environmental changes that would affect multicelled plants and animals. However, we have no record of such changes, and their final effect, being indirect, may appear to be unrelated to the primary cause of which records may exist. The microorganisms provide the infrastructure on which larger species depend, and changes in that infrastructure may play a more important evolutionary role than any with which they have been credited.

Nevertheless, the Precambrian Era apart, the Cretaceous Period is long even on the scale geologists use to measure history, having lasted about 70 million years. The beginning of the period is as far removed from the end of it as the end is removed from our own time. Naturally enough, great changes took place during it. Among those changes were several important extinctions or near extinctions. The marine reptile population altered its composition. The ichthyosaurs—like the one Mary Anning found—were well adapted to life in the sea and yet they seem to have become rarer as the period advanced, and by 65 million years ago they may well have disappeared entirely. The marine crocodiles vanished quite early in the period. However, the plesiosaurs, of still another marine reptile order, flourished and by the end of the period there were at least twenty genera of them. Not all were large but some were—*Kronosaurus* had a skull 3 meters (10 feet) long. The turtles took to a marine way of life, as did some lizards, which evolved into the mosasaurs, also flourishing at the end of the period.

Above all, though, the Cretaceous was the period during which the dinosaurs—land-based reptiles—prospered. When the period began the dinosaurs were already established. They had appeared during the Triassic Period and had become more numerous during the Jurassic, so they had existed for nearly 90 million years before the Cretaceous began. *Iguanodon* was common in the Cretaceous, at least in some places, and the main groups—the sauropods, carnosaurs, coelurosaurs, ornithopods, and pterosaurs—were all present. During the period the sauropods became more numerous but then declined and it is possible they had disappeared before the end of it. The stegosaurs disappeared entirely before the Cretaceous Period began but their place was taken by the ankylosaurs —herbivorous animals whose defense was their armor plating, so that they resembled animated tanks.

These reptiles were dominated by two groups, the ornithischians (herbivores) and the saurischians (carnivores), which together comprise the true dinosaurs. These became diverse and many species became physically large. It was late in the Cretaceous that the largest carnivore the world has ever known appeared, *Tyrannosaurus*—it was not a single species, but a genus with several species and there were many more distant relatives that were rather similar in appearance. The ornithischians produced the hadrosaurs, the duck-billed dinosaurs of which eventually there were some twenty genera, and also the ceratopsians, the reptiles that early in their evolution had large bony frills behind their necks and, by the end of the period, had impressive horns. *Triceratops* belonged to this group. The pterosaurs—the winged reptiles—declined during the period and by its end only a few groups remained.

While it is true that the groups of dinosaurs that thrived produced some giants, the truly immense animals that often spring to mind when the word "dinosaur" is mentioned had disappeared before the Cretaceous began. *Diplodocus,* 26 meters (85 feet) long and truly a giant, flourished in the Upper Jurassic and had disappeared long before the middle of the Cretaceous. *Brachiosaurus,* which grew more than 12 meters (30 feet) tall and weighed 80 tons, became extinct in the middle of the period. *Apatosaurus*—once known as *Brontosaurus*—was a Jurassic species that did not survive long into the Cretaceous.

The picture that emerges during the Cretaceous Period shows the dinosaurs in extremely healthy condition from an evolutionary point of view. Some of them were large—though not all of them by any means—but far from hampering them, their size appears to have given them an advantage. Quite probably the fact that the dinosaurs became first large and then extinct—which happened to several groups of invertebrates at various times—gave rise to the old wives' tale that large size invariably leads to extinction. We must bear in mind that merely because one phenomenon precedes another, it does not necessarily mean there is any causal relationship between the two. As it was, the evolutionary path to great size was followed many times among the ancient reptiles. Many species became extinct, of course. During a period of tens of millions of years we should expect nothing less. Very many mammal species have disappeared during the last 65 million years. We will return to this theme in Chapter 9 when we examine the point that evolution had reached by the end of the Cretaceous.

Before the end of the Cretaceous the dinosaurs were numerous and diverse, and we have no reason at all to suppose them to have been vulnerable to competition except from others of their own kind. They were every bit as successful as are the mammals today. Perhaps they were more successful. After all, they had been thriving for about 150 million years.

In general, species that die out are replaced by other species of the same class. That may sound too sweeping but in fact it is a truism. There are many examples of this from recent history. For one, the aurochs (*Bos primigenius*), a European wild ox, became extinct in the seventeenth century. It was unable to compete with man, who hunted it for its meat and because it damaged crops. The species thus became extinct, but the genus to which it belonged remained healthy—witness its domesticated relatives. There is a long list of species of mammals and birds that have become extinct recently, yet mammals and birds continue to thrive as a class. Our (justified) assumption that particular species may be unable to cope with environmental change must not be extended to an assumption concerning the evolutionary fitness of mammals or birds in general. The two classes are very fit indeed.

The point here is that during the Cretaceous Period, reptiles that died out were replaced by other reptiles, so we have no reason to suppose that reptiles as a class were threatened or that the various dinosaur species were doomed by the ordinary progress of evolution.

It so happens that among the species that have disappeared from Earth in our own century there are the eastern and western barred bandicoots (*Perameles fasciata* and *P. myosura myosura*) and Gaimard's rat kangaroo (*Bettongia gaimardi*), native Australian marsupials that could not tolerate competition from placental mammals introduced into Australia. In their case, because many marsupials suffered severely from placental competition, we may be justified in drawing a conclusion about marsupials as a whole. Just why marsupials find it difficult to compete with placentals is uncertain (the old idea that they are less efficient reproductively probably is untrue), but the fact that they find it difficult remains and may entitle us to conclude that marsupials are less fit evolutionarily than placentals.

In the case of the aurochs, a placental mammal was replaced by other placental mammals—its own domesticated descendants. In the other case, certain marsupials were replaced by species from a different group. During the Cretaceous Period, reptiles that died out were replaced by other reptiles and so we have no reason to suppose that reptiles as a *whole* were threatened or doomed by the ordinary progress of evolution.

At once the gradualist argument seems to weaken, for although many species, of all classes, died out during the Cretaceous Period, this does not alter the fact that many survived to its end—but not beyond its end. It does seem to be true, however, that in modern times, and perhaps since the commencement of the Tertiary, reptiles have continued to lose ground in comparison with other classes. In an evolutionary sense, perhaps they are no longer "strong."

The mammals are known to have existed from the Jurassic Period. At that time they were primitive, but there were several types. However, so far as we know from the fossil record, the animals themselves were small and fairly uncommon. They became

more numerous during the Cretaceous, and by the end of the period the first marsupials and placental insectivores had appeared.
The animals themselves were still small—the largest were about
the size of the modern Virginia opossum or domestic cat. Probably
they were fairly omnivorous, but preferring foods of animal deriva-
tion. No doubt they ate insects, their larvae, and sometimes their
eggs, the eggs of birds, carrion, and such small animals as they
could catch. Even if we suppose them to have been far more
numerous than the fossil record suggests, it is difficult to see what
threat they could have presented to the large, established reptiles.

Despite stories of elephants being terrified of mice (though they
do appear anxious in the presence of small, fast-moving animals,
apparently lest these enter their trunks), populations of large animals are seldom if ever threatened by populations of small ones.
Estimates have been made for the animal population of a modern
European deciduous forest that show that the biomass (the total
weight) of small mammals—rodents, carnivores, and insectivores—
exceeds that of the large mammals—boar and deer—by a ratio of
5:2. In general, boars are omnivorous, while deer are specialized
herbivores. The small mammals are much more diverse. Yet the
small mammals do not inconvenience the large ones. No one has
suggested that the Cretaceous mammalian biomass even approached the reptilian biomass, far less exceeded it, and so it seems
most unlikely that the mammals were serious competitors. There is
no evidence at all to suggest that during the Cretaceous the mammals were proliferating at the rate we should suppose them to do if
they were taking over niches vacated by reptiles. On the contrary,
their proliferation really began in the Tertiary, when they did take
over niches from now-extinct reptiles.

It seems, then, that we must abandon the idea that the decline of
the reptiles was associated in any causal way with the success of
the mammals.

The evidence in respect of birds is more dubious, for here the
fossil record is much less satisfactory than it is for other classes.
The likelihood is, however, that by the end of the Cretaceous modern types of birds had appeared. It is possible that they competed
successfully against the pterosaurs (winged reptiles) and it does
seem that their success occurred at the same time as the decline of
the pterosaurs. (Incidentally, the birds did not evolve from the
pterosaurs, but from a quite different group of dinosaurs.) Even if

the two phenomena are linked causally, that is, if the rise of the birds brought about the decline of the pterosaurs, this does not help the argument, for the birds competed only in the niches available for flying forms. The large land and marine reptiles had never been dependent on food obtained by flying, and so they would be affected by this change no more than modern large ground-feeding mammals would be affected by a change in the population of birds. Furthermore, the decline of the pterosaurs has been observed throughout much of the Cretaceous, and by its end they may have disappeared completely. Yet it is tempting to imagine that they disappeared because they were unable to compete against the birds.

Unfortunately, the argument is not really convincing. We can illustrate the point by comparing birds and bats, an order of mammals. Fossil remains of bats have been found in strata dated as Eocene Epoch of the Tertiary Period—the epoch which began 54 or 60 million years ago—but they may have appeared rather earlier, in the first epoch of the Tertiary Period, the Paleocene. They seem to have caused no great inconvenience to the birds, yet if the success of one group can lead to the elimination of its rivals, they should have done. Bats belong to what may be the most successful of all mammal orders (Chiroptera). They are well adapted to flying and, indeed, can outmaneuver many of the small insectivorous birds with which they share evening airspace.

Although most bats and many birds feed on insects there is no reason to suppose that the emergence of the bats, possibly 60 or 70 million years later than birds, led to the disappearance of a single species of bird. Far less did they cause the disappearance of birds as a whole. Nor can we assume that the nocturnal habit of the bats is a response to competition from diurnal birds. No doubt their nocturnal habit facilitated their establishment, but it is very unlikely that it was adopted for this reason. More probably bats evolved from nocturnal species while birds evolved from diurnal species.

Bats, it is clear, caused no inconvenience to birds, but did birds inconvenience pterosaurs? They may have, but there is no compelling reason to suppose that they did. Many pterosaurs must have fed on insects and many of the larger species probably spent most of their time in the air, rather like the modern wandering albatross or the swift. Their legs were small and weak and would have made it difficult for them to perch, to move on the ground, or to take off again once they landed, although some may have been able to rest on the surface of the water and to take off from there, like modern seabirds.

In the end, we must let the disappearance of the flying reptiles remain a mystery for which we have no satisfactory explanation. All we can say is that the group as a whole disappeared gradually, that probably all of them had gone some time before the end of the Cretaceous, and that their decline coincided with the establishment and subsequent success of the birds.

Nor does it follow that when the decline of one group can be related causally to the success of another the process is necessarily gradual. If we consider the fate of species that have disappeared in modern times as casualties in man's struggle to provide himself with an environment more to his liking, we see that some species have become extinct quite abruptly. In a matter of about three centuries, which to a paleontologist is a period almost too small to measure, more than two hundred species of birds and mammals have disappeared. Although this does not amount to a major extinction comparable with those at the end of the Cretaceous Period, it has been very sudden. And though it has been caused by the very kind of interspecific competition we might suppose would act gradually, it did not, at least not in every case.

It has been suggested that the marine reptiles failed to compete successfully against the more advanced teleosts (the subclass of bony fishes that includes the great majority of modern fishes) that appeared during the Cretaceous. Clearly, the more advanced fishes replaced some—though not all—of their more primitive fish competitors, but it is difficult to see them as an ecological threat to the reptiles. It is more probable that the reptiles would have included the newer fishes in their diet. If it was competition that led to the decline of, for example, the ichthyosaurs, this may be explained more simply, and more plausibly, by the developments that took place among the plesiosaurs, which may have been highly efficient hunters. As with the pterosaurs, we may say of the marine reptiles that disappeared during the Cretaceous only that their disappearance seems to have been gradual, that it was completed before the end of the period, and that it must remain unexplained.

If we can dismiss interspecific competition as an important cause for the species extinctions, at least our consideration of the possibility has yielded a clue that could be important. In our earlier reference to a modern European forest (p. 34) we omitted pointing out a more significant threat to the large mammals: the destruction of their habitat. In this case, man is the agent, but the case is special and the species, or higher taxonomic group, that is responsible need not be, and in fact is unlikely to be, directly in competition with the species that are threatened for the same food or the same type of nesting site.

Where this happens—and it may be the most fully documented of all ecological phenomena—invariably it is the large animals that suffer most severely. Often they are highly specialized feeders and obviously cannot survive the loss of the plant species on which they depend. In addition they lose nesting sites and suffer disturbance they cannot tolerate from the presence of other species, even though those species in fact intend them no harm. Many small animals—especially insects—are also specialized of course, and in this case they usually disappear as well, but their specialized forms are usually modifications of forms that are less, or differently, specialized and which survive. The destruction of tropical rain forest may bring about the extinction of many butterflies and moths, for example, which are totally dependent on the plant communities in which they live, but their disappearance does not imply the disappearance of all butterflies and moths in all parts of the world. Most of the small mammals, however, are highly adaptive. Sometimes they even prosper, as rodents, foxes, and several other groups have prospered by the conversion of temperate forest to farm land.

We must consider, then, the possibility that some major ecological change occurred as a result of which the more highly evolved, more specialized animals found themselves facing circumstances with which they could not cope. We must confine ourselves at this stage to the more gradual kinds of change, for the sudden changes that are implied by some extraterrestrial event are no less ecological or natural.

Such change is likely to manifest itself first as a modification of the trophic base of the ecological system—certain of the microorganisms and the green plants, which convert simple compounds into the more complex substances required by heterotrophs. (Organisms which can manufacture the food they need from simple compounds obtained from the soil, water, or the air are called "autotrophs." Organisms that lack this ability and so must obtain food by eating other organisms are called "heterotrophs.") A change might also occur as a result of events elsewhere in the trophic cycle. For example, the microorganisms which decompose organic wastes and return them to the plant root systems as simple compounds might be altered. Some groups might decline while others whose biochemical activities were only subtly different might prosper. The effect of such change might be experienced in the chemistry of the aqueous soil solution on which green plants, and so all animals, depend, or it might be felt in the composition of gases released into the atmosphere. We know that such changes must have occurred in the past because we have records of their effects. For the moment, though, let us consider only changes among the green plants themselves.

When we think about the course of evolution we tend to think only of the history of animals. Being animals ourselves our interest centers on our own forebears. A lecture on evolution delivered by a tree might adopt a rather different approach! We must not forget that plants have evolved, and continue to evolve, just as animals do. During the Cretaceous the major change in the world's vegetation was marked by the emergence and spread of the flowering plants: the angiosperms.

New plant species may emerge through the normal course of evolution, but it is not impossible to imagine that such new species, once established in particular areas, might spread rapidly into adjacent areas dominated formerly by plants of the older types. All we need do is to suppose some distinct advantage possessed by the new plants and some kind of "ecological leverage" by which they might assert themselves. This is not difficult.

Among the angiosperms themselves we might think first of variations in the biochemical pathways for photosynthesis that have evolved. Principally these variations relate to the extent to which plants use light energy for respiration. The so-called C_4 plants, which do not have photorespiration (they do not use light energy in respiration), photosynthesize more rapidly than those with photorespiration at high temperatures and light intensities and so grow much more rapidly. Such plants are found mainly among the tropical grasses—maize, sorghum, and sugarcane are examples—but since the site of photorespiration is believed to be certain oxidases and catalases found in membrane-bound particles (peroxisomes) within the cytoplasm of plant cells, and since peroxisomes are present in some members of some plant families but not others and even in some species but not others of the same genera—for example *Atriplex* and *Panicum* (saltbush and the grass genus that includes millet and other mainly tropical grasses)—it is possible that they may exist more widely. If so, a climatic warming might well favor them, allowing them to extend their range at the expense of rivals.

There are "gateways" in the lives of plants that could become levers for expansion. Drought may clear away surface vegetation, as it does in the semiarid regions of the world today, and so may prolonged cold or inundation by water. Were species to evolve that could colonize such cleared areas more rapidly than the "traditional" species as soon as conditions ameliorated, then perhaps they could form stands dense enough to exclude their competitors. If so, we need suppose only the existence of such species nearby and a suitable colonizing opportunity, and the vegetation pattern over an affected area could alter radically and very quickly.

Fires, which occur naturally among dry vegetation, remove temporarily almost all plants from the affected surface. Recolonization usually follows in stages until something closely resembling the original community develops. Where fires occur regularly at intervals of a decade or less, the ecological development may be arrested by the success of those species that are most resistant to fire. In parts of the United States, for example, lodgepole pine has formed large stands in this way. The tree itself is composed of a dense, fire-resistant wood, and the seeds, too, are more or less fireproof. There are even tree species which exploit the advantage that permits them to endure fire by producing abundant, and highly flammable, litter. In dry conditions this is liable to ignite spontaneously and the resultant fire eliminates competitors. Is it possible that some such species accelerated the advance of the angiosperms?

To take speculation rather further, is it possible that the proliferation of angiosperms created conditions in which fires were more common? This might happen were the concentration of oxygen in the atmosphere to increase, even very slightly. It might increase if photosynthesis were taking place on a larger scale—during the upsurge of the angiosperms perhaps—and if carbon were being buried rather than being oxidized during the normal course of decomposition. During the formation of coal measures, carbon is buried in this way and at such times the atmospheric concentration of oxygen may well increase. The more oxygen there is, the more readily and more fiercely organic substances will burn.

During part of the Cretaceous the amount of oxygen in the atmosphere did increase. The lush vegetation left records of itself in the form of extensive coal measures. This is convenient paleontologically, and the increase in the concentration of atmospheric oxygen it produced may have constituted a fire hazard.

It is also important climatically. This carbon-burying process prevents oxidative decomposition and so the return of carbon dioxide to the air. Today we worry about the possibility that the burning of the fossil fuels—some of them formed during the Cretaceous—may return that carbon dioxide and lead to a climatic warming through the greenhouse effect (which we describe in Chapter 7). There is no serious reason to doubt that the amount of carbon dioxide in the air has a direct effect on climate. It is possible to calculate the concentration of atmospheric carbon dioxide present at particular times in the past and by examining isotopes of oxygen it is possible to calculate temperatures. Sedimentary rocks contain oxygen compounds from which the necessary samples can be taken, and cores drilled from the rocks can be dated so the age of the samples can be known. The result of such measurements and calculations show that air temperature is directly proportional to the square of the carbon dioxide concentration. In the Cretaceous, therefore, the scavenging of carbon dioxide from the air could have brought about a climatic deterioration. We will return to the effect of climate change later. First we should consider whether a major change in the type of vegetation could affect the animals dependent for food on that vegetation.

At the beginning of the Cretaceous the dominant plants included conifers, ferns, cycads, ginkgos, and other gymnosperms, many of which have disappeared since but which had been well established on land since the latter part of the Paleozoic Era. They became dominant in the Triassic and Jurassic periods and remained so for much of the Cretaceous. In other words, the fauna of the Cretaceous had evolved in a world that supplied them with gymnosperms for food.

The true flowering plants (angiosperms) were present at the beginning of the Cretaceous—they seem to have appeared first in the Mid-, or possibly Early, Jurassic—but they were not common. It was not until the middle of the period that they became more numerous, and by the Late Cretaceous they had come to dominate the flora on all the continents, and they dominate it still. Could the change have happened suddenly?

The principal difference between gymnosperms and angiosperms lies in their methods of reproduction. The gymnosperms produce large, naked ovules and naked seeds; the angiosperms produce minute ovules protected within the ovary of a pistil and seeds contained in a fruit. Quite simply, angiosperms reproduce more efficiently. Whether or not that is the reason for their success, their success is indisputable, and they advanced rapidly from being rare at the start of the period to being only a little more common by the middle of the period and then, in the final 15 or 20 million years, they took over almost all the available plant habitats. The rise of the angiosperms was phenomenal and, in geological terms, it was sudden. Probably their initial advantage permitted them to colonize high ground, but by the end of the period they were present, and then dominant, in the valleys as well.

Could this change have had an adverse effect on animal life? Most certainly it could. We know today that even among the very versatile mammals, the fauna of a coniferous (gymnosperm) forest is markedly different from that of a broadleaf (angiosperm) forest. It requires no great specialization for animals to obtain food from one kind of plant and to find quite different plants unpalatable or actually poisonous. We humans, who evolved among angiosperms, eat angiosperm products, as do our farm animals. The gymnosperms supply us with timber, but with very little that we can eat. Why should we suppose that the herbivorous animals of the Cretaceous were more versatile in this respect than we are? If the change in the vegetation pattern reduced the number of herbivores, the number of carnivores would decline as well, for the obvious reason that their food disappeared.

We know that the vegetation pattern changed and that it changed rapidly. The question we must ask is whether it changed sufficiently rapidly to account for the large number of extinctions that occurred at the very end of the Cretaceous Period. This seems unlikely. Even if this change is the cause, or an important part of a complex cause, it leaves unexplained the disappearance of marine forms that were not dependent on land plants.

The rise of the angiosperms may have been facilitated, at least partly, by changes in climate. On the whole, the Cretaceous was a period of mild climates. If ice existed at all on Earth, it was uncommon. There were no polar icecaps, no glaciers. Between 40 and 90 degrees north latitude, for most of the period mean annual temperatures remained above 10 degrees C. The vegetation of Greenland and Alaska was typical of what we associate with the low temperate latitudes and the subtropics. At the beginning of the period the climate was rather cooler, and mean temperatures fell sharply at the end, but even so they never fell as low as the temperatures to which we are accustomed today. The great periods of glaciation came later. Great forests flourished, in swamps into which vegetation fell and was compressed to form coal measures. This suggests that climates generally were moist.

Conditions, then, were ideal for the production of a luxuriant vegetation and for the success of the animals that fed on that vegetation. The gentle climate was especially suitable for poikilotherms. The temperature never fell so low as to require special measures, such as hibernation or migration, to ensure survival, and if it rose uncomfortably high there was ample shade among the plants and plenty of water in which to bathe.

Having said that, it becomes necessary to qualify it with some care, for the concept of "mean annual temperatures" can be misleading. There is room within it for great local variation, and within such variation it is possible that the climate of particular regions might deteriorate sharply and suddenly while having little effect on world climate as a whole. And there are reasons for supposing that such local effects did occur.

By and large, the continents occupied the same latitudinal positions in the Cretaceous that they occupy today (their drift was mainly east and west) and early in the period there was more dry land than there is now. Little by little, the seas encroached on the land, so that by the Late Cretaceous the total land area was reduced. The mild, moist climate led to the erosion of mountains formed during the Jurassic. New mountains were formed, locally at the beginning of the period, in western North America, Europe, and Japan, and again, and much more dramatically, at the end of the period in the Laramide revolution. This was caused by large crustal movements and it formed the Rocky Mountains, but it was also associated with a general uplifting of land in various places and the formation of other mountain chains. The Atlantic was opening during the Cretaceous so that the character of the seas was changing, too. The most significant changes in the seas were the opening, late in the period, of a long, shallow sea right through North America, linking the Gulf of Mexico with the Arctic Ocean and of another shallow, almost lagoonlike arm of the sea that extended right through Africa and linked the Tethys Sea—the remnant of the ocean that once separated the northern continents from the southern—with the South Atlantic.

Such changes must have affected climates locally. The alteration of ocean currents as new arms opened to link seas that had been separated would have brought cold and warm currents to regions unaccustomed to them, and since the salinity of all seas is not precisely similar, there may have been salinity changes, which might have affected marine species. Even today climatologists cannot calculate reliably and in detail the effect whereby the temperature of the oceans affects air masses and so climates, but no one doubts that such an effect exists and that it is of great importance. If the distribution of warm and cool water altered—though we must remember that in the mild, ice-free Cretaceous the temperature difference between warm and cool water was less than it is now—climatic consequences would follow.

Mountain building would affect climates in three ways. Since it involved raising land above the level it had occupied formerly, the increase in altitude would produce different conditions. The mountains themselves would affect the movement of air masses, creating rain shadows in their lee and more humid conditions on their slopes. The mountain-building process itself would have been associated with increased volcanic activity, and this would have released dust and other small particles into the atmosphere, where they might have been responsible for a cooling, but they are more likely to have brought about a warming. It may be that such changes brought about, or contributed to, the climatic deterioration that marked the Late Cretaceous. As we have seen, however, this deterioration was not serious in itself. The real deterioration occurred much later.

A discussion of the effect on species of a change in climate must not overlook the possible causes of the climate change itself. These are far from irrelevant, for it is possible that the sudden event we propose as a more credible alternative to a gradual process leading to extinctions itself wrought a change in climate—suddenly.

Changing climates are not the only possible cause of habitat deterioration. If the total sea area increased, we may assume that it did so by inundating low-lying land, much of which may have been swampy. This would destroy the habitat for species which preferred the swamps, and as the sea encroached still further into the drier lowlands, yet more habitat would have disappeared.

Did this habitat destruction lead to extinctions? It seems probable that it led to some. Animals that need to spend part of their lives in shallow waters might suffer, but only if we assume the inundation to have been sudden and to have extended rapidly to steeply sloping land, so that the shallow waters were not merely moved gradually from one location to another but removed totally. Perhaps such rapid destruction took place in some places, but in most areas it is more likely that the change was gradual.

The uplifting of land was similarly gradual. We know this from our own experience. The mountains of central Asia, such as the Karakorum, are still being pushed higher as the land to the south rams itself into the main Asian land mass. This mountain-building process has been continuing for millions of years. In geological terms it may well be sudden, but in terms of living organisms it is so gradual as to be beyond observation. The usual response of animals to a slow deterioration in their living conditions is to migrate in search of better conditions elsewhere. Such migrations are not deliberate searches, of course. Individuals and groups of individuals tend to concentrate on the more hospitable parts of their ranges and to extend those ranges where circumstances allow. They will die out only if one area of habitat disappears and no other suitable area is available within their reach. The key is the rate at which the environment changes.

Even then it is difficult to see the process as one capable of producing extinctions among whole groups of species. The destruction of a particular habitat may eliminate species within that habitat, but unless we suppose them to have been very confined geographically, the species will continue to survive in similar habitats elsewhere in the world. We might imagine, for example, that the draining of coastal wetlands in Malaysia would destroy the mangrove forests and thus the species that depend on the conditions found in those forests. Would the mechanism by which the wetlands became dry also dry out coastal wetlands throughout tropical Asia? Would it dry out the coastal wetlands that provide similar environmental conditions in tropical Africa and America? Anything that could render extinct the *Avicennia* mangrove trees would have to operate on such a global scale, producing similar changes everywhere. The most dramatic climate changes of which we know—and not in the Cretaceous, of course—are major glaciations, but even they have not done this. They have eliminated species from large areas in high latitudes, but the species affected have continued to survive, albeit in smaller numbers, at lower latitudes and then recolonized their former regions when the ice retreated.

Rapid change is possible, however. As we saw earlier, it is happening at the present time, as a result mainly of human activity, at a rate some future paleontologist would regard as instantaneous. This paleontologist would also detect, however, the gross environmental changes that brought about the extinctions. In the case of the Cretaceous extinctions we are presented with evidence that at best is ambiguous. Probably there were extinctions due to competition. Certainly there were environmental changes. Certainly the climate changed, even if only slightly at the global level. Yet when we have found plausible explanations for the extinctions that were distributed throughout the Cretaceous, we are left with the seemingly sudden extinctions unexplained.

Finally, though, we must consider the possibility of a gross environmental change, perhaps associated with a climatic change, that was sudden and short-lived. Let us suppose, for example, that the intensity of the solar radiation received at the surface of the Earth were reduced suddenly, say, for a decade or so, or even less. This would cause a great reduction in photosynthesis. As it is, the slight cooling that took place in the northern hemisphere between the 1940s and 1960s reduced the higher-latitude growing season for farm crops by about ten days. If we suppose a more dramatic darkening of the skies, the effect would be increased proportionately. Many plants would die for want of energy, and if the phenomenon were of sufficiently long duration, it would exceed the period of viability of such seeds as were left in the soil. Plants might die out, and most certainly the large herbivorous animals dependent upon those plants would suffer and, through them, the carnivores.

The argument does create a slight difficulty for us here. The modern angiosperms occur throughout the world, but the surviving gymnosperms occur most prolifically in the high-latitude coniferous forests, where winters are long and cold. During glaciations the coniferous forests became established in lower latitudes, sometimes extending as far as the tropics. So perhaps we should expect that any marked reduction in solar intensity would tend to favor the gymnosperms. Yet it is the angiosperms that survived. Further, if the Cretaceous animals were adapted to a predominantly gymnosperm flora, as they must have been, a revival of the gymnosperms should have favored them, too. It should have been the mammals that became extinct!

Perhaps we can rescue ourselves from the difficulty. If the reduction in solar intensity were large but of a limited duration, it is possible first that the angiosperm seeds present in the soil would remain viable and germinate when conditions improved, while the gymnosperms would gain no advantage because temporarily conditions would be impossible for any plant growth at all.

Even so, we are supposing a mechanism by which very large amounts of particulate matter might be placed in the atmosphere. We are supposing, in fact, a major event that was catastrophic in its effects. We have abandoned the gradualist hypothesis.

Gradualist hypotheses that are based on climate changes and consequent ecological disturbances founder ultimately on the fact that the Cretaceous was free from glaciations. The climate may well have deteriorated, but any really important deterioration must involve the change of large amounts of water from the liquid to the solid phase: from liquid to ice. This is a phenomenon with a powerful positive feedback. Snow reflects incoming sunlight and so has a cooling effect. This leads to an increase in the amount of precipitation that falls as snow, which increases the area reflecting sunlight, and the process accelerates. We know a good deal about the effects of recent glaciations, and since this is the most dramatic and most rapid kind of climate change that is known to occur naturally, if any climatic influence is likely to lead to mass extinctions, it should be this. In fact rather few species become extinct, although many are driven from the areas in which they have lived, and many individuals die, because the warmer climates are merely compressed into lower latitudes. Unless we suppose the glaciation to extend all the way to the equator, the tropics will retain sufficient habitat to provide refuges for populations of most species, and when eventually conditions improve again, these will become breeding nuclei from which new habitats will be colonized.

It seems, therefore, that while we may adduce gradualist causes for many extinctions that occurred during the Cretaceous, such causes cannot explain all the extinctions satisfactorily, and most especially they cannot explain the mass extinctions—of many species, remember, not just reptiles—which took place at, and traditionally which mark, the end of the period.

Last of all, the gradualist explanations fail to take account of the chemical anomalies in the boundary clays. In fact, it is not necessary for them to do so, for if a gradualist explanation were to prove acceptable, it is possible to postulate a mechanism to account for the deposition of material in large amounts and from an extraterrestrial source that caused no catastrophe. Were a body to disintegrate in the higher regions of the upper atmosphere, for example, perhaps it could place the requisite amount of material into the atmosphere, from where it would descend, probably quite rapidly, to form a discrete layer on the surface of the land and to sink through the seas to form a similarly discrete layer in the sea-floor sediment.

There is one final, and devastating, blow from which it is difficult for gradualist arguments to recover. When species become extinct we may expect either that the niches they occupied have disappeared entirely so the species are not replaced or that ecological vacancies occur that are filled either by the expansion of existing species or by new speciation. In the case of the Cretaceous extinctions, there was speciation by which the lost species were replaced. The rate at which replacement occurs must parallel, more or less, the rate at which previous species disappeared. We may suppose that this was the case during the Late Cretaceous and early Tertiary. What we see when the evidence is examined is a rapid decline in the number of species at the very end of the Cretaceous followed by an equally rapid increase in the number of new species early in the Tertiary.

This suggests in the strongest way imaginable that the Cretaceous extinctions were extensive, abrupt, and caused by some temporary event that passed, restoring the old niches whose vacation it had wrought and which then were filled. We are looking at the results of a catastrophe.

4

Clouds, Comets, and Meteorites

Have we deluded ourselves? Is it possible that although a great number of species became extinct suddenly at the end of the Cretaceous, the cause of their extinction was a process that was gradual? We must not make the mistake of supposing that gradual causes are incapable of producing sudden effects. Indeed, students of catastrophe theory can produce example after example to prove the contrary.

All of them can be summarized, very unscientifically, by the story of the straw that breaks the camel's back. The point of the story, you will recall, is that the patient beast was loaded with item after item. The weight of its load increased, but gradually by small increments. Finally a point was reached at which the addition of a single straw—the smallest weight of all those that had been strapped to its back—was sufficient to break that back. The gradual process did not change, did not increase in its intensity, yet at a particular point it produced a sudden, typically catastrophic effect. If the example is fanciful, the world offers us many that are not, and the collapse of a bridge as the result of stresses it has born uncomplainingly for years is not dissimilar from the demise of the unfortunate camel. It can and does happen.

Did it happen at the end of the Cretaceous? We believe it did not, that the sudden effect was produced by a sudden cause. Our reason for believing this, however, is not that we are unwilling to attribute sudden effects to gradual causes, but that we can find no gradual mechanism that may have operated at the time to account plausibly for the effect. This would not convince us by itself—after all, we cannot know all that was happening to the world so many millions of years ago. What persuades us to abandon the gradualist hypothesis finally is the apparently greater plausibility of sudden causes.

Nor should we suppose that everyone shares our conviction or even that there is universal agreement among paleontologists that a significant number of extinctions occurred at all at the end of the Cretaceous. There is a little more to be said about that fossil evidence.

Some paleontologists insist that the evidence for sudden extinctions is merely circumstantial and flimsy and this view has been summarized by Thomas J. M. Schopf, one of the world's leading authorities on the subject (in *Science*, Vol. 211, p. 571, 1980), and it is supported by J. David Archibald (*Nature*, Vol. 291, p. 650, 1981). They maintain that in the seas the animals that were affected lived in the upper waters, close to the surface, and that the extinctions generally were felt more severely in the tropics and subtropics than they were in the temperate latitudes. This suggests that the marine species were destroyed by some major environmental change in the surface waters in those latitudes. Many of these species—such as the ammonites—had been in decline for some time, so that the fact that they occur in Upper Cretaceous but not Lower Tertiary strata is of no relevance: they had to disappear eventually and in any case the fact that no ammonite fossil has been found later than the Cretaceous does not prove that no ammonite lived more recently than 65 million years ago. Such ammonites may not have been preserved, or they may have been preserved but in rocks that have not yet been examined. Of course, while the absence of ammonites from Tertiary strata does not prove no ammonites lived later than the Cretaceous, neither does it prove that they did!

In the case of the terrestrial reptiles, it is maintained that the extinctions were confined to the river and floodplain habitats close to the shallow sea that divided North America. There, it is suggested, the sea level fell by some 100 meters. This drained wetland habitats, caused widespread but local changes in the flora, and made the climate more markedly seasonal, factors which combined against animals so large that they needed vast areas of land to supply food for a viable breeding population. We are dealing, therefore, with no more than the disappearance of twenty or so species that in any case were confined to a particular region.

Still the gradualist argument fails to convince. A major environmental change in the upper levels of the tropical and subtropical seas requires explanation, and it appears to have been rather sudden. In fact it accords well with the effect we might expect from a sudden increase in the intensity of ionizing or ultraviolet radiation. (Ionizing radiation is electromagnetic radiation—similar to light or heat—that has sufficent energy to disrupt the structure of atoms, causing them to lose or gain electrons and so acquire a net positive or negative electrical charge. This causes the atoms to react more readily with other atoms, and in the chemistry of living cells this can prove destructive.) This might well destroy phytoplankton that live close to the surface—and probably other members of the plankton (the small plants and animals of the upper layers of the sea and lakes) as well. Anything that kills phytoplankton on a large scale will produce repercussions throughout the aquatic food web and these repercussions will occur very rapidly indeed.

The phytoplankton, after all, are the green plants of aquatic habitats. They feed the herbivores among the zooplankton, which in turn feed carnivores of all kinds in a set of relationships that is far more carnivorous in the seas than it is anywhere on land. The web is completed by the species of the deep ocean, below the level to which light penetrates, whose own food webs are based on nutrients that reach them in the form of matter drifting down from the upper layers of water. Parallel effects might occur among landdwelling species.

An event whose immediate effects were sufficiently dramatic to lead to the extinctions of so many species would produce many secondary effects, one of which may be to alter the balance of oxygen and carbon dioxide in the atmosphere. If the death of the phytoplankton were accompanied by the death of the microorganisms responsible for the decomposition by oxidation of organic matter, or if these organisms survived initially but were overwhelmed by the volume of material presented to them so that their numbers increased until the waters in which they lived became deoxygenated, the result would be the burying of carbon and a consequent increase in the amount of oxygen in the air. This would have climatic consequences, as we have seen.

We must never forget that while we seek solutions for events that affect large multicelled plants and animals, in the final analysis it is the single-celled organisms that regulate environments. They may leave no records, but they are no less important for that, and the effects we can observe may well result from changes among

This two-centimetre thick layer of clay marks the boundary between the Cretaceous and Tertiary periods, 65 million years ago. It separates two beds of marine limestone near Gubbio, Italy. (*Prof. Walter Alvarez*)

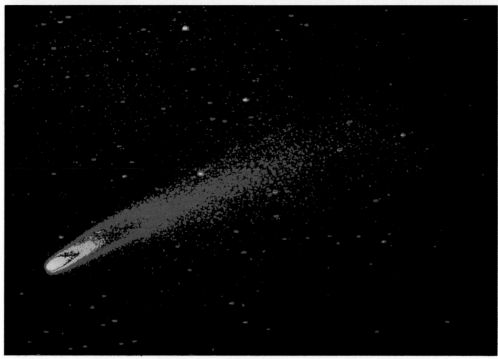

Top. The Crab Nebula is a cloud of gas and dust produced by a supernova explosion in AD 1054, which was observed by Chinese astronomers. It is expanding at a velocity of 1,500 kilometres per second. (*Hale Laboratories*) **Below**. Comet Bennett, seen here in a false-colour photograph, hurtled around the sun in 1970. (*Dr. F. Espenak/Science Photo Library*)

micropopulations brought about by earlier causes we may have overlooked.

If it were true that the large reptiles lived and died out only in part of North America, the gradualist argument would be strengthened—but is it true? According to A. S. Romer (*Vertebrate Paleontology*, 3d ed., Chicago: University of Chicago Press, 1966, considered a principal reference work on the subject), among the Carnosauria whose remains have been found in Upper Cretaceous strata, *Carcharodonsaurus* is known from North Africa; *Chilantaiosaurus* from East Asia; *Dryptosauroides* from South Asia; *Spinosaurus* from North Africa; *Chingkankonsaurus* from East Asia; *Genyodectes* from South America; *Gorgosaurus* from East Asia; *Majungasaurus* from East Asia; *Orthogoniosaurus* from South Asia; *Prodeinodon* from East Asia; *Szechusanosaurus* and *Tarbosaurus* from East Asia; and *Tyrannosaurus* itself from East Asia. Many of these genera, and more besides, are known from North America, but there is no reason to suppose them to have been confined to that continent. We need not trouble to examine other reptilian groups, for the presence of so many of the major carnivores implies the presence of the herbivores on which they preyed.

We must begin, then, to consider the possible candidates for causes of a sudden event. There are several and at various times each has had its adherents. All of them require an extraterrestrial intervention, although one of them seems superficially to be confined to the Earth. We should examine that first.

The inner core of the Earth is believed to be made from solid iron with some nickel and the outer core, which envelops it, largely from iron. The outer core is less dense than the inner core and is believed to be molten—although the difference in the two densities is too great for it to be accounted for only by the different densities of liquid and solid metal. The heat generated within the core causes complex convection movements in the outer core and the rotation of the Earth adds other forces. The precession of the Earth may cause eddying motions in the outer core. If the causes are complex, the motions, of which we can know nothing by direct observation, of course, are more complex still. Their effect is known, however. The inner and outer core together are believed to comprise a kind of self-exciting dynamo which produces a magnetic field. However, the Earth is not a permanent magnet. It cannot be because the temperature within the core exceeds the Curie point—

the temperature at which the magnetic properties of a metal are lost. The field it produces varies greatly over long periods of time both in intensity and in direction, and its intensity at the surface varies from place to place.

When material from the mantle—the region below the crust— reaches the surface and solidifies to form igneous rocks, these rocks record the direction of the Earth's magnetic field at the time of their solidification. Within the mantle, the molten rock is held at temperatures far above its Curie point, which in the case of magmatic rock is quite low. When it reaches the surface of the Earth it cools and solidifies and as the solid rock continues to cool, the Curie point is passed and the atoms within the rock align themselves with the magnetic field in which they find themselves. The rock continues to cool, and as it does so, the freedom of the atoms to move within it is lost. Eventually, in the cold rock, the alignment of the atoms is preserved and so records the alignment of the magnetic field at the time the rock cooled. This alignment can be determined by examination of the rocks, provided, of course, correct allowances are made for crustal movements that may have shifted them from the positions in which they lay at the time they cooled.

Some years ago, geologists investigating rocks in the mid-Atlantic in order to learn more about the way the sea floor is spreading discovered that rocks at different distances from the central ridge had different magnetic alignments. From this it was discovered that during the history of the Earth, at occasional irregular intervals, the polarity of the magnetic field has reversed: magnetic north has become south and south, north. The age of the rocks can be determined radiometrically (by measuring the proportions of certain radioactive elements and the elements into which they decay; this rate is absolutely regular and such dating is very reliable) and so it became possible to date the magnetic-field reversals. Reversals themselves are not instantaneous—the magnet does not "flip"—but they do take place fairly quickly, during a period of one or two thousand years.

Most of the detailed work on magnetic-field reversals has been done using samples taken from rock less than 65 million years old, and so is of little help to us. However, a general picture has been developed of the Earth's magnetic history. This suggests that field reversals tend to occur in batches. There are several within a (geologically) short time of one another, then there is a much longer period during which no reversals occur. It is believed that the Late Cretaceous was a period of reversals, but examination of the boundary clays and rocks to either side of it suggest that no reversal occurred precisely at the end of the Cretaceous.

What might be the consequences at the surface of a magnetic-field reversal? During the period of the reversal itself, the magnetic field would be lost entirely. What must happen at such times requires the field to fade rapidly and then to establish itself again rapidly. Some people supposed that the magnetic field forms a barrier protecting the surface from bombardment by charged cosmic particles—the solar wind—and that in the absence of the field this bombardment might become intense. In fact, though, the effect of the magnetic field—which extends far into space to form the magnetosphere—is more complex, and it is not the magnetosphere that provides most of the protection we enjoy from the solar wind but the ordinary atmosphere—the air that lies far below it. Whether or not a charged particle were deflected by the magnetic field, it would need to be traveling at an almost impossibly high speed in order to reach the surface. Long before it did so, any particle moving at more modest speeds—the speeds actually encountered—would collide with so many gas molecules, ricocheting from one to another, that its power to injure living cells would be lost. On this ground, therefore, it seems the loss of the magnetic field would have little adverse effect on a planet possessing an atmosphere (on a planet without an atmosphere the case would be different).

At one time it was suggested that an effect of the loss of the magnetic field might be to deplete the ozone layer in the upper regions of the atmosphere. This, it was maintained, would permit more intense irradiation of the surface by ultraviolet radiation. Ultraviolet radiation, at the high-frequency end of its spectrum, is held to be dangerous to life because of the damage it can do to living cells. The hypothesis has been abandoned by most people for two reasons.

The first is that as our understanding of the chemistry of the stratosphere has improved, it has become apparent that solar flares are likely to have little effect on the reactions that involve the formation and destruction of ozone.

The second is that the biological effects of ultraviolet radiation have been greatly overstated. We believe, for example, that in the early history of the Earth, before its atmosphere became enriched with oxygen and so before an ozone layer could be formed by reactions involving oxygen, organisms colonized parts of the land surface, and life was abundant in the upper levels of the sea. Ultraviolet radiation is only weakly penetrative, especially in water, but it seems to have caused no great inconvenience to the single-celled organisms living at that time in places that must have exposed them to it, and certain compounds, present in abundance in seawater, are effective absorbers of ultraviolet.

If microorganisms can protect themselves against the injurious effects of ultraviolet radiation, might it not harm larger, multicelled plants and animals nevertheless? Might it, for example, reduce the productivity of marine phytoplankton or of vegetation on land? Frankly, we do not know, but it is possible. Although organisms have "safety margins" which allow them to tolerate changes in their environment that exceed by small amounts the fluctuations to which they are accustomed, there is no reason to suppose they can tolerate gross changes of which they have no evolutionary experience. Thus a slight increase in ultraviolet intensity might cause little serious harm, a larger increase might cause little harm if it were sustained for a very short time, but a large sustained increase might well be damaging. If such damage were caused to plants, we need not consider the direct effect of exposure to the radiation on animals. Whether they survived that or not, they could not escape the consequences of a major change in the amount of food available to them.

We suppose the risk is unreal less because we believe there are no circumstances under which ultraviolet radiation may be injurious than because we are convinced that the mechanism by which the ultraviolet is supposed to reach the surface is implausible: it just does not agree with the way in which stratospheric reactions function, as we understand them at present.

However, we must qualify that statement, for while it is reasonable so long as we suppose the original source of the ultraviolet to be the Sun, if the Sun is not to be the only source, the picture changes. A supernova explosion in our neighborhood in the galaxy, for example, could bathe the Earth in so much radiation that the protective mechanisms at all levels would be overwhelmed.

Is it possible that the magnetic-field reversal was associated with a change in climate? This may be. In 1978 a scientist at the University of Arizona discovered an inverse relationship between the strength of the Earth's magnetic field and the amount of carbon 14 in the atmosphere. This implies an increase in solar radiation, for carbon 14 is formed in the atmosphere by the radiation bombardment of nitrogen. At about the same time, another scientist, at the Lamont-Doherty Geological Observatory, in Alpine, New York, found evidence that the strength of the field is inversely related to temperature: the weaker the field, the warmer the weather. No one suggests the phenomena are related causally, of course. At the end of the Cretaceous, according to the accepted view, the climate was deteriorating. This does not rule out the possibility of a brief period of improved climate—one or two thousand years would be brief on the geological time scale—but it is hardly a reason for the extinction of many species.

We cannot accept a reversal in the magnetic field as a cause for extinctions in itself, but such a reversal may have happened at that time and the two events may be connected. Perhaps there was some event that caused both the extinctions and the magnetic-field reversal.

What might do that? It has been suggested that the polarity of the field might be reversed as a result of the impact of a large body —a meteorite, say—striking the Earth's surface and so causing a major perturbation in the Earth's core. This is quite plausible and it would fit the picture we are building, since a shock sufficiently intense to cause such perturbations would certainly produce highly dramatic effects on the surface!

We must set the large impact to one side for a moment, for there are two other mechanisms by which the magnetic field might be reversed, one of them involving a celestial body, but prior to its impact. Let us suppose that a large body were to enter the atmosphere at very high speed and spinning very fast. This is by no means impossible—we must not suppose it is only the planets that spin on their own axes! As such a body encountered the air, it would ionize atoms of gas with which it collided, and they would ionize atoms on its surface. So it would be surrounded by a spinning cloud of ionized particles: a plasma cloud. This would behave as a kind of superdynamo, to produce a very large—though short-lived, since it would not survive the impact—magnetic field. It is entirely possible that such a violent magnetic shock would cause the Earth's field to alter and perhaps to change polarity.

Were a supernova to explode nearby, this too would produce a huge magnetic shock wave that would disrupt the Earth's magnetic field and quite possibly reverse it.

We conclude, then, that while there is no evidence at present that would lead us to believe that the Earth's magnetic field did reverse its polarity 65 million years ago, the possibility cannot be dismissed. There is no reason to believe that the reversal would produce any adverse effect of itself, and far less is it likely to have caused extinctions. Indeed, it may have been associated with a temporary improvement in the weather. However, while the reversal itself is unlikely to have had any serious biological effect, the cause of the reversal may well have had, since this may have been the passage through the atmosphere of a large object moving at high speed or a supernova explosion nearby. Thus we find ourselves drawn again to an extraterrestrial intrusion.

What might cause such an intrusion and of what might it consist? The most probable candidates are a cloud of dust, a supernova explosion, a close encounter with a large object, or an impact.

We saw earlier that while it is possible for a cloud of dust to produce effects on the surface of the Earth, the cloud that would be required to deposit the layer of noble metals over much of the planet's surface must have certain properties. It would need to be of a certain size and density. No cloud is known in our region of the galaxy that meets the specification. This may be unimportant. Perhaps, since the event, the cloud has moved away and is now hidden from us. Since the solar system travels about the galaxy, perhaps the old cloud is awaiting us in some remote spot and our acquaintance with it will be renewed? Alternatively, its encounter with the Earth may have depleted it to such an extent that now it is much smaller and more diffuse than it was then. We will return to the question of the migrations of the solar system about the galaxy in Chapter 10.

For reasons we will explain in a moment, such a cloud would have to be cold. It cannot have been formed by an event that took place in our vicinity.

In his examination of the boundary clays, Ganapathy found (*Science*, Vol. 209, p. 922, 1980) two isotopes of osmium (osmium 184 and osmium 190) present in proportions characteristic of material originating within the solar system. As we know, osmium can be formed only under conditions found during a supernova explosion.

Ganapathy suggests that when osmium is formed, the proportion of each isotope is characteristic of the particular circumstances surrounding its formation. In other words, the ratio of these two isotopes is typical of a particular supernova event—indeed, it is a signature of that event. Since there is no evident reason why the proportions cannot be different, the chance of two supernovae possessing identical signatures is very remote. Thus he holds that the isotopic ratio in the boundary clays identifies the osmium as having originated in the same supernova explosion as all the other osmium on Earth and, so far as we know, everywhere in the solar system.

His argument is persuasive but it does not amount to a proof. Since we cannot observe the formation of osmium directly, we cannot know that the isotopic ratio is not determined by some mechanism that operates universally. He assumes that the ratio provides us with a supernova signature, but he could be wrong. We should not dismiss the possibility of a supernova cause on this ground alone.

Nor can we dismiss the possibility that a supernova event occurred in our region of the galaxy. Such phenomena are short-lived, but on a galactic time scale they are not uncommon. Five are known to have occurred in our galaxy in the last thousand years. The Crab Nebula is believed to be the result of a supernova event that was observed by Chinese astronomers on July 4, 1054. Within a week, an exploding star may attain a brilliance 10 billion times greater than that of our Sun, and as we saw earlier, the core of the star may then become insignificant. The debris from a supernova remains detectable for some time, by radio astronomy and to a lesser extent because of its emission of visible light and X- and gamma radiation. There are about a hundred such sources in our galaxy. They fade eventually, as the clouds of particles expand and cool (although sometimes they may become heated again, for reasons that are not well understood). Almost certainly, this means that no observable trace will now remain of an event that took place 65 million years ago.

If such an event did occur in our immediate galactic neighborhood, could it have ejected large amounts of material in our direction? Could this have formed a large dust cloud that moved through the solar system, encountering Earth as it passed, and then away into space? Since the cloud is likely to have been highly radioactive, might it have produced dramatic effects? Attractive though the idea is, unfortunately it is implausible. It is true, of course, that the shell cast off during a supernova event consists of particles of dust and gases and that these may form an interstellar cloud. We must remember, though, that the shell is cast off in the form of an expanding sphere—a little like a balloon that is being inflated. It possesses great energy. The shell is ejected at velocities of about 6,000 meters per second. However, the further it moves from the focus of the explosion, the more diffuse it becomes. If it passed through the solar system while it was still sufficiently dense to produce the kinds of effects we require, then the explosion itself would have had to be so close that the initial shock waves would have destroyed everything. Certainly this would account for the extinctions, but it would not permit any survivors at all. The entire planet would have been burned to a cinder! If the supernova were more remote, on the other hand, it might well have produced a more modest effect—enough to destroy many species but not all species—but in this case the cloud it produced would be irrelevant.

This leaves an encounter with a solid body as the remaining suspect we have not discussed. We may leave until later the question of whether such a body struck the surface of the Earth directly or whether it merely grazed past us. We should establish the ability of the suspect to commit the crime before we move on to a consideration of the modus operandi. There are two classes of candidates: comets and meteorites. We must treat them with caution, for there are dangers in proposing either of them as the cause of a major catastrophe. We have preconceptions about them that have accumulated over millennia in the cultures of every civilization on Earth.

Because many of them are bright, spectacular, and visible for days on end, comets have always interested people, and at various times they have been blamed for bringing plagues, the failure of crops, and other disasters. Similarly, meteorites at various times have been accused of being the thunderbolts that are hurled by Zeus, Jupiter, Thor, Perkunas, Perun, Taranis, Yahweh, and many less renowned gods to punish wrongdoers. Perhaps the charge is unjust. The true thunderbolt was a magical, invisible weapon that caused the physical damage seen after lightning has struck. All the same, we must take care that our rational argument does not contribute to mere myth.

As a matter of fact, and superstitious idiots though we may like to believe our forefathers to have been, in the matter of thunder-bolts they were less wrongheaded than they may seem since thun-derbolts do exist. They are produced during very violent thunder-storms, when cumulonimbus clouds extend all the way up to the stratosphere. Under such conditions it is possible for a lightning flash to spark from the very top of the cloud to the ground, with the discharge, at the ground, of so much energy that soil is fused into lumps of glassy material. Obviously, no one sees them form at the surface and equally obviously they were not there before the lightning flash, and so it is not entirely unreasonable for people to imagine that they hurtle from the sky to the ground. For the rec-ord, they occur most commonly in the temperate latitudes in spring and autumn.

Comets are objects of uncertain origin that move within the solar system in orbits that carry them alternately close to the Sun and very far away from it. Each time a comet passes close to the Sun it loses some of the material of which it is composed, and there is no way in which it can make good such loss. In time, therefore, most comets disappear by attrition, and some are lost from the solar sys-tem entirely by gravitational disturbances in their orbits that hurl them away into interstellar space. Because of this attrition and be-cause of the ways in which their orbits and speeds relative to the Sun change, comets themselves alter their characters and behavior.

It was the American astronomer Fred L. Whipple, of the Smith-sonian Astrophysical Observatory, in Cambridge, Massachusetts, of whom we shall hear again in Chapter 6, who published a paper in 1950 in which he proposed that, contrary to the theory prevailing at that time, comets possess solid nuclei, made from about equal proportions of rock and crystals of frozen liquids and gases. In fact, they are, in his phrase, "dirty snowballs." This idea is now accepted by most scientists. Their rocky portions consist of boulders and many smaller fragments, with a total diameter of about 1 kilome-ter, all held together in the ice, which increases their size. The mass of a comet is much less than that of a completely solid meteorite of similar size.

This would be important if the speed of the comet relative to the Earth were much the same as the speed of any other body in the solar system, but we cannot be sure that this is so. Once away from the Earth we have to be rather careful what we mean when we use words like "speed" because, of course, this is relative to the observer who measures it. The Earth moves in its orbit at 29.8 kilometers per second. The speed of the comet Bradfield has been calculated as 24 kilometers per second relative to the Sun and that of the Kohoutek and West classes of comets as about 50 kilometers per second. If these were also their speeds relative to the Earth, then indeed a comet would need to be larger than a lump of solid rock to achieve the same effect as the lump of solid rock on impact, because its speed would not be greatly different from that of a meteorite. However, cometary orbits are highly irregular and it is quite possible for one to encounter the Earth head on. In this case the two speeds—of the comet and of Earth—must be added together, and at an impact speed of 50 to 80 kilometers per second the gap between the effect achievable by a comet and that achievable by a meteorite narrows considerably.

Kinetic energy, which is what matters, is calculated by multiplying half the mass of the body by the square of its velocity. We see, therefore, that if we compare a lump of solid rock traveling at 20 kilometers per second with a comet whose mass is only one tenth that of the solid lump, it will possess the same kinetic energy if its velocity is increased to a little over 63 kilometers per second. (The arithmetic goes like this: let the mass of the lump of rock be 100 and that of the comet 10; 100 divided by 2 and multiplied by the square of 20 is 20,000; 10 divided by 2 and multiplied by the square of 63 is 19,845.) The numbers are not entirely fanciful, for the average comet has a diameter of about 1.25 kilometers. Because of the difference in density between a "dirty snowball" and a meteorite, a comet with the mass of a meteorite 10 kilometers in diameter would have a diameter of about 12.5 kilometers; if half the mass of the comet is ice and can be discounted, leaving us with only the denser rock portion to consider, then we must double its volume to give it the mass it would have were it entirely solid. The volume of a sphere doubles with an increase of about 25 percent in its diameter, so a comet would be 12.5 kilometers across if it had the same mass as a meteorite 10 kilometers across. It is unlikely, though not impossible, that a comet would be as large as 12.5 kilometers, but a velocity of 63 kilometers per second for a comet with a diameter of 1.25 kilometers is well within the observed range. Indeed, the velocity might well be higher, so that the size of the comet could be smaller in proportion.

The difficulty with the idea of a cometary impact lies not in any inherent implausibility, for it is entirely plausible, but in the chemistry of the boundary clays. It seems unlikely that a "dirty snowball" could deliver the requisite amounts of osmium and iridium.

We considered in Chapter 2 the possibility that a comet could have released poisons into the air or the sea and found that unlikely, too. The reason is quite simple. Let us suppose a comet with a nucleus 1.25 kilometers in diameter. This would give the nucleus a volume of about 1 cubic kilometer. Half of that volume, or 0.5 cubic kilometer, consists of solid substances that would be liquids or gases at the temperatures found at the Earth's surface—what we would call "ice," meaning a substance that is found only occasionally in the solid state. We must remember that on Venus, for example, solid lead might be considered "ice" because it would exist there normally in the liquid state. Let us allow that all of this, the entire 0.5 cubic kilometer, consists of cyanides and other poisons. Let us suppose, further, that it fell into the North Atlantic (a supposition for which we have reasons that will emerge in the next chapter). At the end of the Cretaceous the North Atlantic was much narrower than it is now, but probably its mean depth was much the same. It was very roughly triangular in shape. Its dimensions seem to have been about 3,000 kilometers wide at the base, about 3,000 kilometers from north to south, and probably about 4 kilometers deep. This gives it a volume of 18 million cubic kilometers. We are supposing, then, that 0.5 cubic kilometer of toxic substance fell into 18 million cubic kilometers of water. Living organisms in the vicinity of the impact would have been killed by the impact: they would not have survived long enough to be poisoned! Poisoning would have affected only those organisms that were some distance from the impact, but as the poisons approached them they would be dispersing in the extremely turbulent conditions into a mass 36 million times greater than their own. Some organisms may have been poisoned. It seems unlikely that many could have been, and it is much less likely that so small a quantity of poison, diluted in the oceans of the world—which included the Pacific, which was much larger and deeper than the Atlantic—could have delivered lethal or even mildly harmful doses to the great majority of organisms. It could not have caused the extinctions, and even if it did, we have not accounted for the extinctions of land-dwelling species.

We are left with meteorites and asteroids. There is a slight confusion of terms here, since according to one theory the asteroid belt is believed to be the source of the meteorites that move about the solar system and occasionally collide with planets. An alternative view holds them to be the remains of cometary nuclei—so explaining their irregular orbits. Meteorites are bodies that collide with planets and asteroids are similar bodies that do not. Those meteorites whose orbit carries them into the vicinity of the Earth are known as Apollo asteroids. Nineteen of them are known to exist. They have diameters of between 1 and 6 kilometers, except for two of them (known as P-L 6344 and 1976UA) which may be no more than 200 meters in diameter, but some of them approach very close to Earth and it is not impossible that they might cross between the Earth and the Moon. Those that exist now are too small to account for the event we are seeking to explain, but there may have been larger members of the group in the past. Obviously, their collision with planets would remove them from the group. In composition we may take it that they are typical of other meteorites.

There are two main types of meteorites, those made from rock and those with a metallic (mainly iron) core—the siderophiles. The chemical compositions of the two types are different and while either could have caused a major impact event, the metals deposited in the boundary clays are characteristic of those found in siderophile meteorites.

It seems most probable, therefore, that only one class of objects could have caused major effects at the surface and at the same time could have deposited the metals found in the boundary clays. We are considering an iron meteorite. However, since the provenance of the body remains unknown, perhaps it is best not to call it either a meteorite or an asteroid. Instead, from this point on, we shall use the less precise, more general, term. We shall call the object a planetesimal, a tiny planet.

Next we must try to picture what the encounter or impact must have been like.

5

The Barrel of Fire

The sea blew up. Transformed into incandescent gas, the ocean leaped upward with a roar that would have been heard on the other side of the world.

There could have been no warning. At least, there could have been no warning that might have helped any of the animals living on Earth at that time. Perhaps the more alert of them, the very quick-witted, might have caught a glimpse of a flash in the sky, might have begun to raise their heads. Were such reptilian heads protruding above the surface of the ocean? Were the seabirds—which existed by then and looked much like their modern descendants—whirling and screaming in their incessant search for food? Did they begin to notice the fire in the heavens? It is doubtful, and the observation would not have helped them.

This is not because the event would not have been visible. Far from it. When the approaching planetesimal entered the outermost regions of the atmosphere, some 150 kilometers above the surface, it would have begun to shine more brightly than the Sun. Because of the angle it subtended, it would have appeared some 10 times larger than the Sun and as its temperature rose to some 18,000 degrees C—3 times hotter than the Sun—it would have become 100 times brighter. No more than a second after it appeared it would have filled the sky. The observation of it would have been no more than a glimpse because any observer close enough to see it would have been burned to a cinder by the radiant heat that preceded it. The sea immediately below the object would have begun to boil vigorously shortly before the impact.

For living organisms in its path, the end would have been quick. A planetesimal falling almost vertically to the surface at 20 kilometers a second would take less than 2 seconds to pass through the denser region of the atmosphere. Some comets may travel at up to 4 times that speed in relation to the Earth, and they would take proportionately less time to arrive. For countless millions of animals, all sleeping, grazing, stalking of prey, quarreling, courting, mating, would become oblivion. They would cease to exist.

The explosion that tore the ocean apart would have been more violent than anything we have known in our history, more violent than anything we can imagine without help. To try to comprehend it we must break it down, to describe each of its appalling elements, one at a time.

In the first place we must ask you to accept our assertion that the object fell into the sea. It makes little difference to many of its immediate effects whether it fell on sea or on land, but we have reasons for supposing that it fell somewhere in the middle of the North Atlantic, and we will give them later, in Chapter 6.

Let us begin with the size of the planetesimal. It had a diameter of 10 or 11 kilometers. This sounds little enough, but that is because we are comparing it with other members of the solar system. It is hardly large enough to qualify the body for the dignified title of "planet." It is much the same size as Phobos, one of the small, lumpy satellites of Mars. Phobos is about 13.5 kilometers long and 9.5 kilometers wide. Our Moon is much larger. The planetesimal is small on the cosmic scale, but that is the wrong scale. We should compare it with objects on the Earth itself.

Mount Everest is just over 8,850 meters high. That is to say, the highest peak of Everest (although it may still be growing higher because the crustal movements that formed the Himalayas have not ceased) stands about 8.9 kilometers above sea level. If you imagine Everest standing on a column of rock, isolated so you can see it from summit to sea level as it were, and if you imagine the planetesimal set upon the ground beside it—without sinking into the depression caused by its great weight—the planetesimal would be 1 or 2 kilometers taller than Everest, and much wider. It would dwarf the mountain.

It would present a hazard to aviation. Only military aircraft and intercontinental civil airliners fly at altitudes of 10 kilometers or more. A Concorde would clear it comfortably, but a Boeing 747 or a Tristar would have to make a detour around it unless the captain was prepared to entertain his passengers with a little low flying.

We are dealing, then, with an object that is small on the cosmic

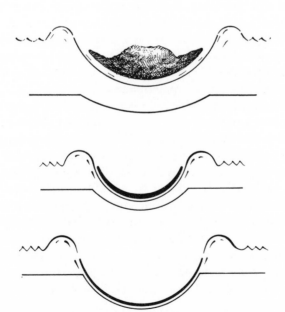

This is a reconstruction of the likely behaviour of a large object as it strikes the sea and sea bed. Note how the object itself is deformed, so that no recognizable body is left behind.

scale but very large indeed on any scale that is familiar to us on Earth. It was also very solid. It was a lump of rock no whit less solid than Everest itself, and it can have been no less dense, so 1 cubic meter of it would have weighed no less than 1 cubic meter of rock from Everest. The Alvarez team has estimated its weight at between 100 billion and 1,000 billion tons. This allows plenty of room for refinement (Ganapathy estimated 2,500 billion tons), but at least it gives an idea of the mass of the thing.

It approached the Earth at about 20 kilometers per second (nearly 45,000 miles per hour). The Earth itself moves in its orbit around the Sun at 29.8 kilometers per second, but the numbers mean little. Twenty kilometers per second is almost 60 times the speed of sound (Mach 1) at sea level in the atmosphere. At Mach 60 the journey by plane from London to New York would take a fraction more than 4 minutes. The airline would save money on catering and lose it in bar sales and workaholic executives would find themselves facing the 29-hour day of which they pretend to dream. The *Columbia* space shuttle orbits the Earth at about 8 kilometers per second (about 18,000 miles per hour). A satellite in geosynchronous orbit, whose position remains constant above a point on the Earth's surface, travels at about 3 kilometers per second (6,700 miles per hour). An M16 rifle can fire a bullet at speeds of very nearly 1 kilometer per second (2,200 miles per hour). Since the speed of a high-velocity rifle bullet is something to which we can relate, let us use that. We are considering an object much larger and heavier than Mount Everest, made from solid rock and metal, approaching the Earth at about 20 times the speed of a high-velocity bullet from a modern army rifle.

It was big and fast, and its impact released a great deal of energy. How much energy? We can calculate that because we have estimates for the mass and velocity of the planetesimal. Again, though, we produce a number—in this case about 1,000 billion ergs per square centimeter of surface—that means nothing. W. H. McCrea has related this number to something more familiar (*Proceedings of the Royal Society*, Vol. 375, No. 1760, pp. 33–34, 1980). He estimates that the impact would have been equivalent to the detonation of 100 trillion (100 million million) tons of TNT. How big is that? How big is a number 1 followed by fourteen zeros? The atomic bomb dropped on Nagasaki in August 1945 exploded with a force equal to about 20,000 tons of TNT: a 2 and four zeros. The

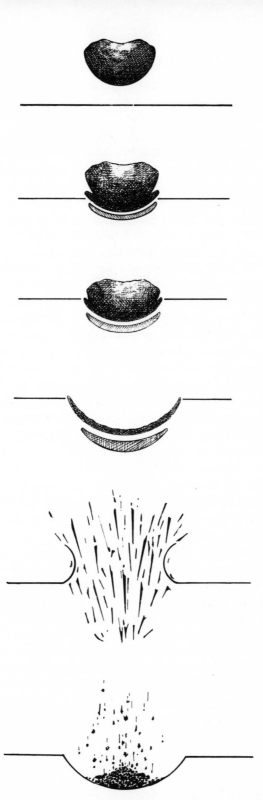

This is a reconstruction of the likely behaviour of a large object as it makes its impact on land. Note the rise in the centre of the crater, while the impacting body disintegrates and leaves little of itself behind in the crater area.

planetesimal therefore arrived with as much energy as about 5 billion (5,000 million) Nagasaki-sized bombs. To help us further, McCrea points out that had the energy been distributed evenly over the surface of the Earth (which, fortunately, it was not) it would have amounted to about ten Nagasaki-sized bombs on each and every square kilometer of the planet's surface: about twenty-six Nagasaki bombs per square mile. If we suppose that the entire global armory of nuclear weapons can be measured as a few hundred billion tons of TNT equivalent, then we are trying to imagine the simultaneous explosion, in one place, of about 1,000 times that entire stockpile. It is hardly surprising that we find it difficult! We are attempting to describe a phenomenon that is far beyond anything we have ever experienced or that is recorded in our history. We must point out, however, that while we may speak of the total energy released by the impact, the effects of that impact depend greatly on the way the energy was released. As we shall explain later, the huge amount of energy released by the planetesimal was far more diffuse than, and so quite unlike, a nuclear explosion.

Nevertheless, the scene we describe is so far removed from anything that has been experienced by human beings, so far removed even from any phenomena contrived in laboratories, that we can do no more than speculate about the physical, chemical, and biological implications. What we or any others say or write about the detail of the event must contain a very large amount of informed guesswork.

There is one particular natural disaster that happened about a century ago and that is well documented. The eruption of the volcano Krakatoa, in what is now Indonesia, on August 27, 1883, is said to have been the most violent volcanic eruption ever recorded and perhaps the worst disaster of its kind in recorded history. The tsunamis ("tidal" waves) alone killed some 36,000 people on the neighboring islands of Sumatra and Java. Probably Krakatoa exploded with the energy of something like 1,000 million tons of TNT. Our planetesimal impact released energy equivalent to about 100,000 Krakatoa-sized eruptions.

What happens when 100,000 Krakatoas explode all together? We should not exaggerate the force involved. It was nowhere near sufficient to disturb the orbit of the Earth—our world was not knocked out of place. Nor was the explosion anything like big enough to cause severe structural damage to the planet itself. To a large extent this is due to the relative plasticity of Earth: even its solid, hard rocks can bend and flow just a little and the entire crust moves constantly and often crumples. There are other bodies in the solar system—the remote, deep frozen satellites of Saturn, for example—that are much more brittle. They can shatter under a strong impact, and in the past probably they have shattered and reformed as the fragments fell toward one another gravitationally. In order to bend or shatter a surface, however, the impacting body must reach that surface. Did it?

Some people have supposed that a body entering the atmosphere at such a high velocity would not survive long enough to reach the Earth's surface. It would burn or disintegrate under the stresses caused by its passage through increasingly dense air, releasing all of its energy into the atmosphere. Had the planetesimal done this— had the explosion been of an air-burst type—the consequences would have been more severe because the energy would have been delivered over a wider area of the surface. It is unlikely that its passage through the air had any serious effect on the planetesimal at all. It is true, of course, that most very small bodies which enter the atmosphere are destroyed by heat. We see them sometimes as "shooting stars." These objects, which delight or terrify us, range in size from dust particles to lumps of matter about the size of a baked bean. Even so, some of them reach the surface, along with meteorites ranging in size from that of small pebbles to tennis balls to—happily rarely—that of boulders weighing a ton or more. These bigger objects are too large to waste away by attrition in the atmosphere, but still small enough to be slowed greatly in their fall through the air, so although they enter at high speed they are traveling fairly slowly by the time they reach the ground. The extent of the heating depends on the speed of the body and the length of time it spends in the air—which depends in turn on the angle at which it enters the atmosphere. For the moment we are supposing that the planetesimal approached the surface nearly vertically, that it spent 2 seconds or less in transit through the atmosphere, and in this time, although it would have heated at the surface and may well have lost material, the loss would have been quite insignificant as a proportion of its total mass.

It is the encounter with the sea that would have demolished the planetesimal. Again, it is tempting to suppose that a body that could penetrate the atmosphere almost as though the atmosphere were not there could penetrate water with little more difficulty. As anyone will confirm who has ever misjudged a dive and belly flopped into a swimming pool, there are occasions on which water may behave in a breathtakingly solid fashion. Airmen who bale out over water are advised to enter the water in an upright stance, feet first (how they achieve this stance is less clear), and if they can point their toes, like dancers, so much the better for their personal comfort.

Now, falling airmen—even those whom unkind fate has separated from their parachutes—and inadvertent belly floppers travel quite slowly. Their problem arises because they present a large surface area of themselves to the water and the water is unable to move in an accommodating manner to receive them. In order to yield gently, water must move aside as the body enters it. Obviously, the smaller or more gradual the movement of water that is required, the more easily that movement is achieved. The more graceful diver, whose body is streamlined so that it enters the water fingertips first, experiences no pain on impact because only a small movement of water is needed to permit the entry of first the fingers, then the body that is attached to them, a little at a time.

Why should the water move at all? When a solid body moves through air or water, the air or water it displaces must go somewhere—it cannot simply cease to exist. In fact the air or water in front of the body is pushed aside and it closes again behind it. This pushing of the fluid medium creates pressure waves which move ahead of the body and start the parting of the medium before the body arrives. The body travels, as it were, cocooned behind the pressure waves its movement creates. If the body is slender—like an arrow—the displacement needed to permit its passage is small. When pressure waves in air encounter a mass of water, the waves continue to be propagated in the water, but because the water is 1,000 times more dense than the air, a greater expenditure of energy is required to move it. If the body is slender, all may be well, because the energy of motion of the body—its kinetic energy—may be sufficient. If the body presents a large surface to the water, so requiring a large displacement, its kinetic energy may be insufficient, the water may be "unprepared" and the result is sore skin—or worse. Our planetesimal may have been almost any shape, but it is most unlikely to have resembled an arrow. It would have belly flopped, and the difference on impact between

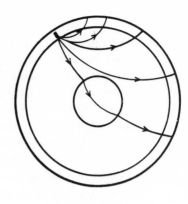

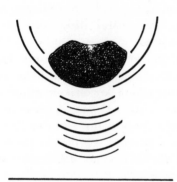

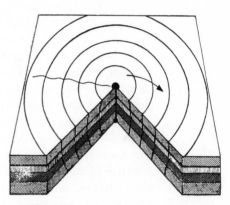

These two cross-sections of the earth show how seismic waves travel out from the point of impact

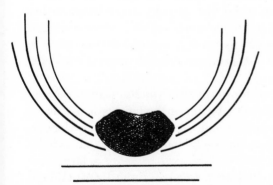

Shock waves from an object as it approaches and passes through the speed of sound. Note how the shock waves are squeezed back towards the object as its speed increases until finally they trail behind as the object passes through the "sound barrier".

water that was no longer yielding and rock and metal ore would have been of academic interest only. What is more, in this case there is another factor we must take into account.

Pressure waves travel ahead of a moving body at the speed of sound. This means that the distance they extend in front of the body depends on the speed of the body in relation to the speed of sound. As this speed approaches that of sound, the body moves closer to its own forward pressure waves and at the speed of sound, and at any speed in excess of the speed of sound, it generates no forward pressure waves at all. The fluid through which it moves receives no advance warning. The waves are still formed, but the body overtakes and travels ahead of them and they are compressed very tightly, into the sonic boom that rattles windows and sets householders trembling with indignation.

Sonic booms are caused by the the compressed pressure waves emitted by the moving body in the shape of a cone behind the body, and the waves—and so the sound—become more diffuse with increasing distance from the body. If the body—say, an aircraft—moves horizontally, the cone of pressure waves will meet the ground surface, tracing an invisible but audible path whose width will depend on the height of the aircraft itself. The bang we hear is due to the pressure waves near the edge of the cone, where they are most intense because they are nearest to the aircraft causing them. If the aircraft were moving vertically the experience would be different. If it was descending there would be no sound at all until after it hit the ground, then the pressure waves would arrive not as a single bang, but as a sustained roar. Similarly, if it were climbing away from us we would hear only the roar, and no sonic bang at all, even though the aircraft itself exceeded the speed of sound.

The planetesimal, descending vertically, would have produced pressure waves that would have begun to arrive, as a sustained roar, immediately after the impact and would have continued to arrive for as many seconds as the body had taken in its journey through the atmosphere.

Back in the 1940s, when high-performance aircraft encountered the so-called sound barrier for the first time, some people believed that this barrier would prove insuperable. When an aircraft reached it, went the argument, the air would behave as a solid through which no aircraft could pass. This proved incorrect, of course, and aircraft can travel faster than sound, but the idea seemed reasonable enough at the time. The real problem turned out to be the turbulence of the airflow over the control surfaces on the wings and tails of aircraft as they approached close to the speed of sound and the consequent loss of control. It was solved by a combination of aircraft design which ensures that the laminar flow of air over control surfaces is maintained at transsonic speeds and engines more powerful than any that existed in the early 1940s. So far no man-made object has moved through water faster than sound. The sound barrier in water would be a much more formidable obstacle than its aerial cousin.

The planetesimal, then, would have encountered water that was not warned of its approach and so behaved in a somewhat solid manner, and then to compound the body's misfortunes it would have belly flopped into it. It is unlikely to have survived the experience. We know that large meteorites that crash to the Earth on dry land cause a crater, but a crater which contains no relic of themselves. A similar body entering water would suffer a similar fate, but again, we must remind you that we are dealing with an event that is entirely unfamiliar to us.

It would have made a prodigious splash. The splash, of course, is the aquatic equivalent of the crater-forming process we might expect on dry land, and we can make a reasonable guess at the sequence of events that would occupy the last few milliseconds of the planetesimal's existence.

As the leading edge of the planetesimal met the unforgiving water, its movement would have been slowed, but the remainder of the body would have continued to travel at something very close to its original speed. The shock of the impact would have been transmitted back through the body, but the waves transmitting it would have moved at the speed of sound. The speed of sound within the body would be much greater than the speed of sound in air, but since we suppose the body to have been moving at 60 times the speed of sound in air, it is probable that not only would the water have received no advance pressure waves, but neither would most of the interior of the body.

If we imagine that on the far, uppermost side there sat an observer who leaped aboard the planetesimal somewhere in deep space in order to obtain a free ride to Earth, and if we assume that he found tolerable the heating of the outer layers of his cold lump of rock to a temperature 3 times that of the Sun, we see that he suffered no great inconvenience upon the arrival of his own antipodes. It is not until he reached the shock wave itself that he was discomfited by being vaporized along with what remained of his cosmic steed. This terminal event would have occurred approximately half a second after the lower edge struck the water and would have left him insufficient time to record a descriptive message for our benefit.

The body would have deformed as the lower edge was flattened and pressed into a hollow formed in the water. The water, you will recall, would have been boiling vigorously as the body approached, and the surface of the body itself would have been at a temperature of some 18,000 degrees C. The pressures and temperatures within the band where rock and water made contact would have caused rock and water alike to be dissociated into their constituent atoms and the atoms to be ionized—stripped of their electrons—to form a cloud of plasma. The characteristics and behavior of such a plasma cloud are so markedly different from those of more familiar states of matter that we must consider this stage of the event as though three masses were involved rather than two. We have the solid body that so far remains unaffected, the water that remains so far unaffected, and the plasma formed where they meet. The plasma is a gas, but a gas of great density, and it would be expelled to the sides, away from the center of the impact. A substantial proportion of the body would be lost before it encountered the sea bed.

The overall effect would be rather like that achieved when a leaden air-gun pellet is fired at a brick wall. The pellet splatters, but it also makes a little dent in the wall.

Let us think, then, about this dent—the crater. We are supposing that the planetesimal fell into the North Atlantic. Sixty-five million years ago the North Atlantic was much narrower than it is today. Probably it was about 3,000 kilometers wide at its widest point, and rather triangular in shape. In the north the continents were still joined. We cannot know how deep it was, but since it was widening then and is widening still, because two tectonic plates are moving apart and material from the mantle is filling the gap, it is reasonable to suppose the ancient Atlantic was of much the same depth as the modern Atlantic, and that averages something like 4 kilometers. We know, too, that an object striking the Earth with the energy we calculate this object possessed would have made a crater on dry land about 200 kilometers in diameter and about 20 to 30 kilometers deep. In the sea the depth of the crater would have been rather less, because the rock of which the oceanic crust is composed is denser than that of the continental crust. The upper part of the crater would have been formed in water, however.

The water and rock would have formed an intensely hot, intensely dense fireball that expanded very rapidly sideways and upward, exposing the ocean floor over most of the 200-kilometer diameter. If the planetesimal had been spinning when it entered the atmosphere, the plasma fireball might also have been spinning and would have been a superdynamo producing a strong magnetic field. The fireball would rise, not because it was diffuse—like ordinarily hot air—but because there was no other direction in which it could move. Some 1,250 billion tons of water would be removed and carried into the air.

Walter Alvarez has calculated that the amount of material ejected in an impact of this kind is equal to about 60 times the mass of the impacting body. In this case, therefore, we must suppose that between 6,000 and 60,000 billion tons of matter entered the atmosphere, most of it from the seabed. On the seabed, allowing for the depth of water and for the greater density of oceanic as compared to continental crust, the crater might well have been 3 or more kilometers deep, and temporarily it might have been much deeper.

The amount of water may seem impressive. Indeed, it is impressive. We must remember, though, that the oceans form but the merest film on the surface of our planet. If the Earth were the size of an orange, the oceans would comprise no more than a slight film of moisture over part of the skin. The crater itself would have formed a minor pit in the skin—not much different from the pores that pit the skin of a real orange. This comparison helps to explain not only the scale of the effect on the surface of the Earth, but also its shape. The material ejected from the crater as a hot, dense plasma, would have been almost disk-shaped. It would have risen into the sky like a very wide barrel: a barrel of fire.

It would have risen very fast indeed, like a flash. It is not impossible that it would have attained escape velocity and that at least part of it would have vanished into space. More of it may have risen high enough to enter an orbit about the Earth. Would it have formed a disk, like the disks of Saturn? It might, if it carried fragments of solid material all of different sizes. If such fragments were to enter stable orbits, their orbital speeds would have been different, each according to its mass, and most probably they would have orbited at different heights. This would have tended to separate them until they formed a ring. On the other hand, plasma material would reform as a fine dust and it might orbit as a cloud.

As the sea leaped to the sky, water from the surrounding ocean would have rushed in to take its place. The crater would have filled with water in a few minutes, a mighty rush of water from all sides producing pressure waves of its own which would have interfered with one another as the circumference of the incoming waves became smaller. As the inrushing waters met at the center they would have risen as a huge, secondary spout, this time of water rather than plasma. What would happen when cold water encountered molten rock? We do not know with any certainty, but we can be reasonably confident that the secondary spout would have contained large amounts of superheated water vapor.

The heat of the impact would have caused much of the ejected material to rise high into the atmosphere, creating an area of extremely low atmospheric pressure beneath. This would have filled, but how would it have filled? Had the rising air and water been at temperatures with which we are familiar, we could speculate about the meteorology of the event, but a fast-rising cloud of plasma is something that defies our imaginations.

We have a picture, then, of tremendous turbulence, a great swirling of air and water, over the ocean itself and extending to the coasts on either side; of a huge roar, with the din from the descending planetesimal concealed in the cacophony; of a light much brighter than the Sun. We may imagine an immense cloud, beside which the notorious hydrogen-bomb mushroom is but a puff of smoke, mounted on a stem more than 200 kilometers across, and extending tens or hundreds of kilometers upward to where the plume spread to the sides as though against an invisible ceiling. The cloud would have been so large as to have been invisible as such from the Earth itself, and much of it would have been masked by the secondary cloud of water vapor and dust trapped below the tropopause, the boundary above which air temperature no longer falls as altitude increases. From the Moon, perhaps, it would have appeared as a cloud. To a terrestrial observer far enough from the

site of the impact to escape immediate destruction, the entire hori-
zon would have seemed engulfed in fire, and as the fire rose—faster
than the keenest eye could follow—the whole sky would be illumi-
nated by the fire barrel.

We have a picture. No sooner have we built this picture in our
minds than we must erase it, for as the plume of debris spread
across the whole sky, the world round about was plunged into
darkness.

The drama of the impact must have been visible over a wide
area, but eventually the curvature of the Earth would have
screened it from the gaze of more distant eyes. They would have
seen neither the impact itself nor even the glow reflected from the
base of the clouds that might have given warning that all was not
well with the world. Before long, though, such warnings would
have begun to arrive.

The impact would have sent seismic waves through the Earth itself and these would have provided the first warning, for they
travel faster than sound waves in water or air.

There are several types of seismic wave, each with its own characteristic velocity. Primary (P) waves are compressional waves
that move parallel to the direction of propagation. Secondary (S)
waves are shear waves that move more or less at right angles to the
direction of propagation of pressure. Because the speed of waves
increases as the medium through which they are propagated becomes more dense, both kinds of waves follow curved paths
through the Earth. P waves move at speeds of a little more than
6,000 meters per second (about 20,000 feet per second) to about
10,400 meters per second (34,000 feet per second). S waves travel
at about 3,400 meters per second (11,000 feet per second) to about
7,200 meters per second (23,600 feet per second). Because they
travel faster, P waves always arrive ahead of S waves. There are
also surface waves. These are propagated parallel to the surface of
the Earth, rather than moving down through it, and they travel at
around 4,000 meters per second (13,000 feet per second). Seismic
P and S waves may reach the boundary of the Earth's core, from
which they are reflected, and they can also be refracted, much as
light can, so that the sequence of signals reaching a seismic station
monitoring an earthquake is very complex. Here we deal only with
general ideas, and the picture we paint is much simplified.

Consider the experience of an observer positioned as far from the site of the impact as it is possible to be—on the far side of the Earth exactly half an Earth circumference away. The first P waves would have arrived a little more than 20 minutes after the impact that caused them. The S waves would have followed and the surface waves begin to arrive a little more than 80 minutes after the impact. The Earth would have trembled perceptibly but at so great a distance probably not violently. Trees might have shaken but it is unlikely that they would have fallen. The earthquake would have consisted not of a single tremor, but of a rumbling that would continue for an hour or more as wave after wave arrived. The surface waves might have been more violent than they would be from an ordinary earthquake, whose focus is much deeper below the surface than the planetesimal impact. It is possible that although their energy would have been dissipated by the movement of the rocks through which they traveled, they might have had sufficient energy to make more than one circuit of the Earth. If so, the tremors would have begun about 20 minutes after the impact, reached their peak during the hour that followed, then ceased for about 2½ hours, by which time the surface waves would have been arriving on their second circuit. They would have arrived from all directions and it seems likely that the turbulence caused by their meeting might have caused a small "mirror image" of the impact shock at the point on the Earth precisely opposite the focus of the disturbance.

The next set of shock waves to arrive would have traveled by water and so they would not have been perceived by an observer on land and out of sight of the sea. The speed of sound in fresh water at room temperature is about 1,500 meters per second (3,355.57 miles per hour). In cold seawater it is perhaps 1,000 meters per second (2,237 miles per hour), and at this speed the waves would have reached the opposite side of the world in a little more than 10 hours. These sound waves would have had little immediate effect that was visible to an observer on land. The surface of the sea would have gone white as a layer of spray formed just above it: the same effect is visible as the first warning that a depth charge has exploded below the surface of the ocean. In the water itself the effect on marine animals would have been catastrophic. Unable to tolerate sudden violent changes in pressure, many fish would have been crushed. For days afterward their bodies would have floated on the surface, food for scavengers but ignored by those predators whose killing and feeding behavior was stimulated only by the presence of live prey. Many modern reptiles are predators of this kind. The only consolation is that such waves attenuate rapidly. Since the North Atlantic was roughly triangular in shape they would have moved only in a southerly direction and most probably their power to kill all but the most sensitive of organisms would have been lost before they reached the equator. The shock might have been detectable at greater distances, visible even, but it would have been fairly harmless.

Then the sound would have arrived. Traveling at about 344 meters per second (769.5 miles per hour) through the air the sound waves would have taken rather more than 16 hours to reach the opposite side of the world.

At least, the sound *might* have arrived. If it did, the sound waves, traveling by sea and by air, would not have arrived as a single "bang," but as a deep, rumbling roar that might have continued for some time. It is not impossible that, like the surface waves moving through the crust, the sound waves made more than one circuit of the planet, so that the sound would return several times, like echoes of the crack of doom. Sound waves attenuate rapidly in air, the extent of the attenuation—or loss—depending on air temperature but also on the frequency of the sound itself. The high frequencies are lost most rapidly, which is why distant sounds are most commonly heard as low, rumbling noises. In this case our observer would have been likely to hear a very deep rumbling indeed, for in traveling halfway around the Earth, sound with a frequency of 1 kilohertz (kHz)—1,000 cycles per second—will be reduced by 76,000 decibels in warm air and a sound at 100 hertz (Hz) by only 6,000 decibels. Sound at 20 Hz will lose no more than 200 decibels, but this frequency is below the limit of detection for the human ear.

Last of all, the tsunamis would have arrived—the tidal waves that have nothing whatever to do with the tides—and they would have swept inland, devastating everything in their path and leaving behind desolation and a soil sterilized by its soaking in salt water.

Tsunamis are caused by seismic shocks generated below—or in this case at—the seabed. In fact they are gravity waves and invariably they are caused by a splash, usually made by a sudden movement at the seabed. They are just big waves, like those we make in a bathtub. They travel slowly, at no more than 178 to 220 meters per second (398 to 492 miles per hour), and in the open, deep sea they have wavelengths of a kilometer or more and a wave height at the surface of no more than 1 meter and often less. Their speed and wavelength are related: the greater the wavelength, the greater the speed. They travel rather slowly, but their progress is much faster than that of ordinary waves—or tides—and sailors seldom notice them at all. They rush past, amid the ordinary turmoil of the sea, and if a ship is buffeted, the buffeting feels like nothing more than one slight bump in a world of bumps. It is when they reach shallow water that their behavior changes. The name "tsunami" is a Japanese word that means "harbor wave." Constrained by the shelving seabed, the length of the wave is reduced, its amplitude increases, and it is slowed—and consequently piles up—so that the wave rises to form a wall of water that may be traveling at hundreds of kilometers an hour, and so far as living organisms are concerned, that is fast enough to make escape difficult.

Most tsunamis are no more than 3 or 4 meters high, despite their fearsome reputation, but much larger ones are known. Those following the 1883 eruption of Krakatoa, which devastated coastal areas to either side of the Sunda Strait, some 15 kilometers from Krakatoa, were 38.4 meters high and others caused by the eruption were recorded on the west coast of South America and in Hawaii, although at such great distances they caused no damage.

But we are dealing now with a seismic event far more violent than the Krakatoa eruption. Certainly tsunamis would have occurred on oceanic coasts everywhere in the world, and on the North American and Eurasian coasts, closest to the impact site, the tsunamis are likely to have been hundreds of meters and maybe kilometers high.

The entire world, therefore, will have trembled as seismic waves moved from the focus of the event through the rocks of the Earth. Shock waves moving through the water would have followed. Then the sound would have arrived—the blast and its echoes. Finally, tsunamis would have pounded every exposed coast. But these are only the immediate, first effects. There would have been others, and we must consider them.

6

Sky, Sea, Rock

As the immediate shock effects died away, the secondary effects would have become apparent. These can have been only a little less dramatic than the immediate effects and they were far more important from an evolutionary point of view.

The shock, felt throughout the crust, might have caused secondary earth movements, leading to more earthquakes and an increase in volcanic activity. Volcanoes which were about to erupt or which would have erupted anyway in a few years might well have erupted at once. It is unlikely that new volcanoes would have appeared or that long dead volcanoes would have been jolted into life. Volcanoes are the outward manifestations of pressures which build up very slowly, over long periods of time, as material collects in a magma chamber some distance below the surface. The material itself—the magma—is under very great pressure so that while it behaves as a liquid in many respects, while it remains beneath the surface it is an extremely dense, viscous liquid. It is most unlikely that any impact at the surface, however large, could accelerate this process.

Fred Whipple, whom we mentioned earlier as the man who first described cometary nuclei as "dirty snowballs," has suggested (in *New Scientist*, Vol. 89, No. 1245, p. 740, 1981) that a body 10 kilometers in diameter entering the North Atlantic at 20 kilometers per second would puncture the oceanic crust, making a hole about 100 kilometers in diameter. He believes that the impact site may have been on or very close to the mid-oceanic ridge and that the tectonic activity it triggered led to the emergence of Iceland.

Perhaps we should say something here about the structure of the outer layers of the solid part of the Earth. It is believed that the uppermost rocks form rigid plates—tectonic plates—which move on the surface of a mantle composed of material that behaves like a liquid. The picture is graphic, but it can be misleading. In the first place, the mantle material is more dense than the rigid plates which lie above it. Movements within it are very slow. Heat moves through it by convection, as in more familiar liquids, and if the great pressure under which it is held should be released—as in a volcanic eruption—it may flow as a liquid, but it is quite different from what we usually understand by the word "liquid." It is not in the least like water, for example. Perhaps, then, we should find a replacement for the word "crust," because the crust of the Earth is not at all like the crust on a pie, a solid cover above a liquid.

There is a clear boundary between the crust and the mantle, called the "Mohorovičić discontinuity," or "Moho," after the Croatian geophysicist Andrija Mohorovičić who discovered it, but we must remember that it is a boundary between upper material that is dense and lower material that is very much more dense. When Whipple speaks of "puncturing the crust," therefore, we must not picture this as we might the puncturing of a piecrust or of surface ice on a pond. It is rather the removal of upper material to a depth sufficient to expose the denser, hotter material beneath. Had this happened, there is no doubt at all but that the encounter between hot magma and cold or even boiling seawater would have produced an immense second explosion, as the water rushed back to replace that which had been removed in the cloud of plasma.

In any event, Iceland sits squarely on top of the mid-oceanic ridge, on top of a linear volcano which produces geysers, hot springs, and boiling mud to prove it. It is also true that Iceland continues to grow—the emergence from the ocean off the coast of Iceland of the island of Surtsey some twenty years ago, proved that to a startled world. Iceland is, of course, intensely interesting to geologists—because it lies above a linear volcano—and has been studied very thoroughly. No rocks more than 65 million years old have been found on it. Whether or not the impact of a planetesimal was the cause or contributed to the cause, as Whipple claims, it is a fact that Iceland emerged from the sea following a large volcanic eruption about 65 million years ago. So although at first glance Whipple's idea may seem extravagant, it is not. Many large impact craters have mountains at their centers—the Moon is covered with

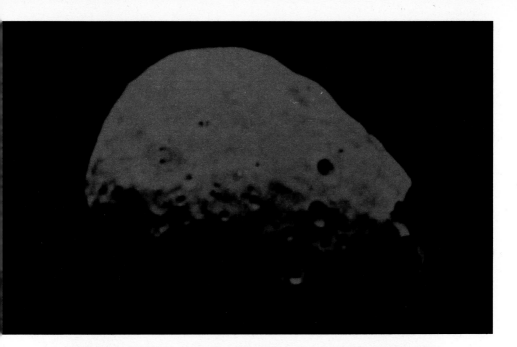

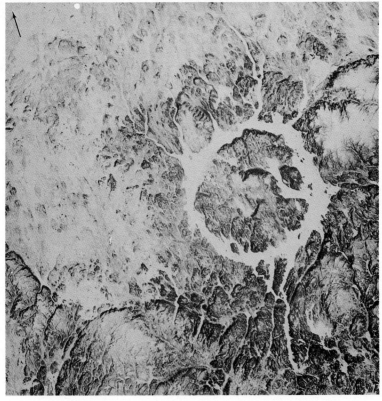

Top. One of the moons of Mars, Phobos, is a potato-shaped planetisimal approximately 22 kilometres in diameter. It is seen here in a Viking spacecraft photo taken in July 1976. (*NASA/Science Photo Library*) **Below.** The Manicougan Lake in Quebec, seen here in wintry guise by the Landsat spacecraft, is a crater 66 kilometres in diameter. It is probably caused by the impact of a giant meteorite. (*NASA/Science Photo Library*)

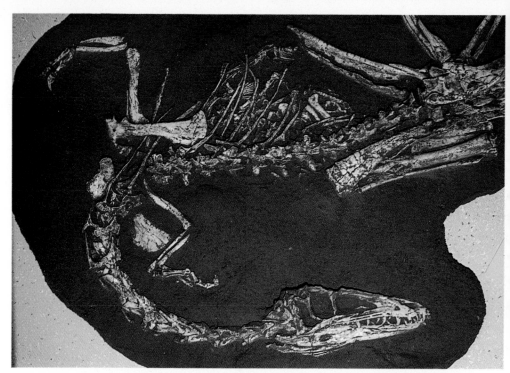

Top. This fossilised skeleton belongs to a Coelurosaur, a small bird-like dinosaur which was amongst those that perished in the Great Extinction 65 million years ago. (*Pat Morris*) **Below.** The Ammonites flourished in the world's oceans for some 375 million years before the Great Extinction. Molluscs with a many-chambered spiral shell, they were amongst the many other types of organism which were annihilated along with the dinosaurs. (*Sinclair Stammers/ Science Photo Library*)

craters of this kind. Dare we suppose that volcanic Iceland is the mountain at the center of such a terrestrial crater? No one knows, of course, but the possibility is intriguing.

It is possible that the planetesimal would have punctured the Earth's crust. The oceanic crust is made from rocks that are much denser than those which form the continental crust, so it is probable that a seabed crater would be markedly shallower than a continental crater. This difference in rock densities between continental and oceanic crust was discovered when the gravitational force over both types was measured. It was known already that oceanic crust is much thinner than continental crust. In places the continental crust is some 40 kilometers (25 miles) thick, but the average thickness of oceanic crust is a mere 7 kilometers (4 miles). If the rocks were of similar densities, therefore, gravity should be just a little stronger over the continents than it is over the sea floor. In fact it is not, and this can mean only that 7 kilometers of oceanic crust is as dense as 40 kilometers of continental crust, ergo, the oceanic crust is made from rocks 5.7 times more dense than those which form continental crust. If an impact of the kind we are considering were to form a crater from 20 to 30 kilometers (12 to 19 miles) deep in continental crust, first we must deduct the depth of water, then take account of the difference in rock density. We find that the equivalent of a 20–30-kilometer continental crater on oceanic crust lying beneath 4 kilometers (2.5 miles) of water is a crater between 2.8 and about 4.5 kilometers (1.7 and 2.8 miles) deep. This is the final crater, however, and the physical effect of the impact would have been felt at much greater depths, probably at least 10 kilometers (6 miles). This means that while the planetesimal itself would not have penetrated the crustal rocks—little of it would remain by that time—it might well have released mantle material by partly vaporizing the overlying rocks.

The impact must have made a formidable dent in the surface. Let us forget for the moment the suggestion that Iceland may lie at the center of the crater and look for the crater itself. Where is it now? Surely, we should be able to find some trace of so catastrophic an event, something more substantial than a little cosmic pollution in the boundary clays.

There is no denying that the discovery of a crater of the appropriate size and age, in a plausible location, would provide much support for the impact theory. But its absence does not seriously weaken the theory.

The crater may exist but may not have been found; indeed, we may never find it. If the planetesimal fell into the deep ocean—as we suggest it did—its crater might be difficult to detect. The ocean floor is more thoroughly mapped than it was just a few years ago, but the maps have many blanks and filling them is a slow, expensive business. Our insatiable appetite for minerals and the likelihood that such minerals may be found in rich hoards on the bed of the oceans encourages treasure-seeking mining corporations to accelerate efforts to explore the seabed and pay for the mapping, but even so, it will be years before oceans are mapped as clearly as continents. In any case, the feature we seek may be well hidden.

The spreading of the sea floor in the North Atlantic, which has been proceeding since long before the end of the Cretaceous, moves the continents apart, but when we speak of continents in this context we include the parts of them that are submerged beneath the ocean: the continental shelves that are the low-lying edges of the continents themselves. Compared with the deep bed of the mid-ocean, the shelves are submerged beneath shallow water. They are fed by rivers that carry silt to them and there is no reason to suppose that 65 million years ago they were smaller in area or less deeply covered in sediments derived mainly from the land than they are today. If this is so, the water which rushed in to fill the crater following the impact would have brought with it great quantities of such sand and mud scoured from other parts of the seabed, and in the 65 million years that have elapsed since then more, much more, sediment may have arrived. If the crater lies in the ocean deeps, therefore, it may well be undiscovered, and if it lies in shallower water on a continental shelf it may be buried beneath a deep layer of sediment.

Had the planetesimal fallen on land, as the Alvarez team believes it did, no trace of the crater would probably exist today. Its rim would have been eroded away by the weather, its basin filled, the landscape flattened, and, quite possibly, twisted and folded by subsequent movements of the crust. The fact is that the crust of Earth is too mobile to provide a good surface on which the history of such accidental encounters could be recorded indelibly.

This is not to say that a crater on land may not exist. Many impact craters do, and some of them are both large and ancient. They are not easy to detect, however, many of them betraying their origins only when the landscapes in which they occur are photographed from space. Then they may be revealed as vast circular landforms hidden from ground-based observers—beneath dense vegetation, for example. It is rather doubtful whether a crater 200 or more kilometers across could be detected from the ground in, say, the South American tropical forest, and even in the Canadian tundra it took satellites to reveal the presence of an impact crater whose existence was previously unknown.

The movement of rocks has another effect. As we saw earlier, the outer solid surface of the Earth consists not of a single layer of material like a skin but of a number of plates of hard rock that float on the surface of denser but more plastic matter. The plates move, and because they move, yet provide a continuous surface to the planet—there are no holes in the crust—they alter the surface continually. In some places, plates slide, more or less jerkily, past one another. In other places, they move together and one plate is crumpled upward to form mountains. In still other places, one plate rides over the top of its neighbor, so that a boundary is formed at which a plate is being subducted—depressed and pushed under—into the mantle, where its rocks are melted and any records they may bear are destroyed forever.

It is reasonable to suppose that, given sufficient time, all of the material that forms the crust of the Earth will be subducted into the mantle, its place being taken by new material. It seems this has been the case throughout the history of the planet, for no part of the oceanic crust has been found that is more than 200 million years old. Over a period of 65 million years there is about an even chance that a feature anywhere on the bed of an ocean which lies above a destructive plate margin will be removed to the mantle by subduction. (This does not apply to shallow seas, such as the Mediterranean or North Sea, which form in depressions in the continental, not oceanic, crust.)

As we mentioned earlier, the North Atlantic has been growing wider since before the end of the Cretaceous. Its widening is caused by the movement of two plates that are drifting apart. As they move, the space between them is being filled by new material from the mantle, so that new seabed is being formed all the time. There is no region of subduction in the North Atlantic, and no part of the crust is being destroyed there. The process sounds gentle, but magma does not reach the surface gently. The eruption of mantle material is, by definition, a volcanic event. Were a crater to be formed across the region where the plates join, the movement of the plates would divide the crater into two parts, its walls would possibly be demolished, at least partially, by volcanic eruptions, and, very probably, the depression would be filled to some extent with new rock. Today the walls of a crater that once were 200 kilometers apart would therefore be 1,000 kilometers apart or more, and, if those walls still exist, indistinguishable from minor submarine mountain chains.

We have asked you to accept, temporarily, our assertion that the planetesimal fell more or less vertically and that it landed in the North Atlantic. It is time now for us to defend that assertion.

Why do we suppose that the body fell steeply? After all, it could have entered the atmosphere on a somewhat tangential path, approaching the surface in a more shallow descent. Could it have made more than one circuit of the Earth before it crashed into the surface?

In the first place, we must ask you to consider the geometry of the approach. The best way to begin may be for you to imagine a full Moon as it appears from Earth, and consider the problem of hitting it with a missile from a gun. The Moon is a mere 400,000 or so kilometers away and it offers a hemispheric target similar to that which the Earth presents from a greater distance to any approaching body. The angle through which you may move your gunsight between one edge and the other is very small. You will be very lucky to hit the Moon at all, and if you do hit it, the chance that you will score a bull's-eye on its center, or anywhere near its center, is very small indeed. It is much more likely that you will score an "outer" and hit the target somewhere near the edge. Because the target is three-dimensional rather than two-dimensional, this would mean that the approach was at a fairly shallow angle in relation to the surface.

The diameter of the Earth is about 13,000 kilometers (8,000 miles), and therefore the surface area of one hemisphere is about 1 billion square kilometers (50 million square miles), which sounds large enough for even the most incompetent marksman. The target measures a little more than 20,000 kilometers (12,000 miles) from edge to edge. Not all of this exposed surface can be counted, however, because an object approaching at high speed very far to the edge most probably would miss the target altogether. So perhaps we may allow a distance of, say, 8,000 kilometers (5,000 miles) from the center of the hemisphere to the outer limit a body could in practice strike. The critical factor is the length of time the missile takes in its passage through the atmosphere, and now we can calculate this. The vertical depth of the atmosphere is about 150 kilometers (93 miles). Let this be the height of a right-angled triangle, and let the base of the triangle be 8,000 kilometers. If the missile travels along the hypotenuse of this triangle, it will move through about 8,000 kilometers of air, and at a speed of 20 kilometers per second this will take it almost 7 minutes, rather than the 7.5 seconds it would take were its approach vertical.

As it moves through the air, friction caused by air molecules will cause it to heat and to lose energy, and the missile may disintegrate. The release of so much energy into the air would devastate the atmosphere, heating it strongly, causing chemical reactions within it, and releasing substances into it. It is difficult to see how any living thing could have survived an air burst of this kind. However, the extinction was not total: there were survivors. It is the fact of the survivors that leads us to suppose that less of the missile's energy was released in the atmosphere itself and therefore that the time spent in passing through the atmosphere was less and the angle steeper.

What, though, if by some chance the body were to be thrown into a spiraling orbit about the Earth so that its angle of descent were reduced to just a few degrees and its speed was also reduced so it behaved like a returning spacecraft? Its speed would have to be reduced, because it would be quite impossible for any body to possess a velocity of 20 kilometers per second in an Earth orbit. At this velocity, it would fly out of orbit and away into space. Therefore, if we suppose it entered an orbit which decayed to bring it to the surface slowly we must devise some way in which it could be slowed before it reached the vicinity of Earth. We are quite unable to do so! If it had happened despite this, there could have been one of two results. Either—and this is the more probable—the body would have disintegrated while at the very outer limit of the atmosphere, so it would have formed a cloud of debris, some of which might have fallen in later; or it would have been slowed to such an extent that it fell intact. In that case, it would still exist as a rather large lump of solid matter.

Our picture of the immediate effects of the impact are based on our calculation of the energy the body possessed. Regardless of the angle at which it approached, it must have possessed the same amount of energy, because this is determined by its mass and velocity. In its passage through the atmosphere and the ocean and into the crust of the Earth, it loses energy until, at last, all of the energy is dissipated. If we suppose it to have entered at a very shallow angle, we are supposing that it spent a longer time traveling through the air, in which case it would have lost much more of its energy to the air itself. It would have been heated by friction much more thoroughly. Perhaps it would have begun to break into large fragments. Its remains would have showered down over a very wide area. Who knows but that it might have ended its existence not with a single explosion, but as a belt of debris spread right around the world. In this case, we would need to look no further for the crater, since such an arrival would have made no crater.

It is not impossible that a comet, made largely from ice, might arrive in this way and we have some idea of what would happen if it did, for the Tunguska region of Siberia (see Chapter 2) may have experienced such an event in 1908. Instead of one large crater, many small craters, 50 to 200 meters in diameter, were found in the vicinity of the impact, suggesting that the body disintegrated in the air. All animal life within 1,000 square kilometers was killed by heat. The comet that some believe caused this devastation would have had a mass of only about 50,000 tons. Clearly our much larger planetesimal would have had a much larger effect.

Almost all of our planetesimal's energy would have been dissipated in the air itself. It would have heated the air, as well as been heated by it. Such strong heating—and remember we are considering the energy of 100 trillion tons of TNT, whether this is released on impact or not—would have made life intolerable not just for those species that became extinct, but for every living thing. Vegetation would have burned, large animals would have cooked—literally—and microorganisms would have been parched in their sterilized habitats.

Nor is this all. Small fires radiate a certain amount of heat, but their effect on material around them is caused mainly by convection or by the comparatively slow heating that makes combustible material smolder, then begin to burn. Such a fire may seem to spread rapidly, but it is possible for fire to spread even more rapidly. When the energy of the fire exceeds a certain threshold, the radiated heat becomes so intense that material some distance from the center ignites, adding to the amount of heat radiation, and it is this radiation which propagates the fire. The result is that the fire spreads in a flash—explosively. There have been such fires. During World War II the intensive incendiary bombing of cities occasionally produced this appalling effect. It is very likely that it was such a radiation-propagated event that caused the severe burning at Tunguska, and if a large planetesimal were to deliver its energy—a surface temperature some 3 times that of the Sun, remember—into the atmosphere it is impossible that anything could have survived. Indeed, the very atmosphere itself might have been destroyed.

It was not destroyed—we are here to prove it. True, the effect on living organisms was calamitous, but there were survivors. Life continued, and so we must limit the catastrophe to one that, though serious, is survivable.

There is a further difficulty with the theory of the shallow descent. Even if we imagine that the body entered with much less energy, so that it did not explode violently in the air (as might be the case if a comet were to approach the Earth on a path that was almost parallel to the Earth's own orbit and at a slow relative speed), it would not have distributed enough of the metals iridium and osmium in the pattern in which we seem to find them. A long approach, in the course of which the planetesimal becomes more or less fragmented so that it reaches the surface without a large impact, would deliver the matter of which it is made in discrete areas, perhaps confined to a particular locality or perhaps in a belt around the world or across a continent or ocean, but it is difficult to see how it would have distributed itself over the entire globe—which it did. Even if it could have, for it to supply the quantities of metals we know exist demands that we revise our ideas of the composition of comets. However, this may be less unreasonable than it may seem. It is a point to which we shall return in Chapter 10.

We conclude, therefore, that the body approached the Earth

steeply, that it disintegrated on impact, and that it formed a crater even though we have found no trace of the crater.

Why do we suppose it fell in the North Atlantic? We must return to the distribution of the noble metals in the boundary clays. In the Pacific Deep Sea Drilling Project cores, you will remember, the boundary clays were significantly enriched. In New Zealand they were enriched rather more, and in Spain and Italy they were still more enriched. In Denmark, however, they were enriched to an enormous extent. In fact, while the New Zealand clays contain these metals at about 20 times the average crustal abundance and the Italian clays contain about 30 times more, the Danish clays contain *160 times* more. Our picture of what happened must explain this anomaly: we believe that a fall into the North Atlantic does so.

Let us return to the great cloud of plasma that we left rising above the site of the impact. What would have happened to it? What would have happened to the inrushing water as it encountered molten rock? We cannot know what might have happened under such conditions—we cannot even guess. So far as we know, no one has studied the meteorology of plasma clouds! The meteorology of ordinary terrestrial air is quite complex enough without complicating it further with magnetohydrodynamics! We would need to discover, for example, what the effect might be if gases were drawn toward an area of low pressure around which the Coriolis force (the force at right angles to the direction of movement, caused by the rotation of the Earth) would try to twist them in one direction while a strong magnetic field tried to twist them in the other direction. Anything could happen. The only thing of which we can be certain is that whatever did happen, its consequences for living organisms would not be enjoyable.

Let us suppose, however, that a portion of the material from the planetesimal itself was ejected into an orbit around the Earth, that a second portion was held in the upper atmosphere, and that a third portion remained in the troposphere (see p. 118). Clearly, the energy required to convey these three parcels to their destinations will be greater the further they must go, and so it seems to follow that the most energetic parcel was the one which entered orbit with the plasma cloud or its remains. What happened to all this material? That which was in orbit would have descended slowly, over a

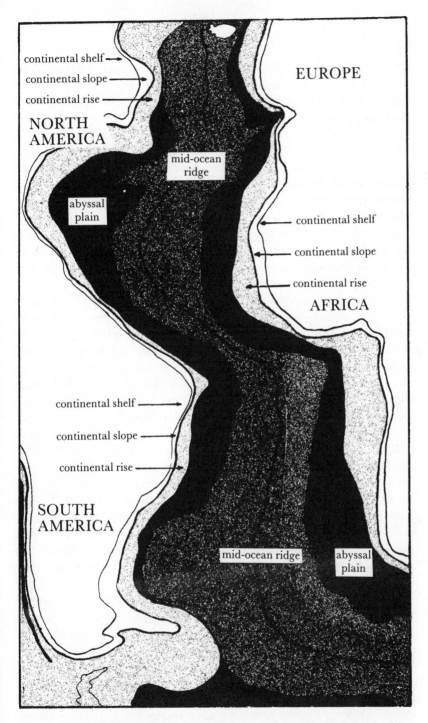

continental shelf →
continental slope →
continental rise →

NORTH AMERICA

EUROPE

mid-ocean ridge

abyssal plain

continental shelf ←
continental slope ←
continental rise ←

AFRICA

SOUTH AMERICA

continental shelf →
continental slope →
continental rise →

mid-ocean ridge

abyssal plain

This is the bed of the Atlantic Ocean as it is today, with the Atlantic ridge running down the centre straddled by Iceland in the north.

period of years, and, depending on the way it was distributed in orbit—as a cloud enveloping the entire Earth or as a ring, for example—it would have been distributed either as a fairly even layer over the entire surface or as a belt around the Earth. If we allow that some of the material entered orbit, then we must agree that it descended, for it is not there now. The material in the upper atmosphere would also have descended gradually, but ahead of the orbital material. It would have been distributed evenly around the Earth, though perhaps contained within discrete latitudinal belts. The material in the lower atmosphere would have fallen most quickly of all, and our knowledge of fallout—from industrial chimney stacks and from atmospheric nuclear explosions as well as from volcanoes—leads us to suppose that most of it would fall not far from the point at which it was released into the air. The density of the fallout decreases rapidly with distance from the impact area. What is more, the less energetic material in the lower atmosphere would cool fairly quickly to temperatures at which the water vapor within it condensed. Because of the difficulty we have with the arithmetic of plasma meteorology, we cannot be confident about any description of the condition of the air over the ocean following the rise of the fire barrel, but we do know that it returned to normal, probably within a short period, and a good deal of the dust in that dusty air must have been washed out by rain.

It is not too misleading to think of the event as something like the firing of a shotgun: the shot is propelled from the gun, but there is also a muzzle flash, a sheet of flame that consists of the hot combustion products from the explosion of the gunpowder. In our case, the muzzle flash represents the plasma cloud as it leaped, quite literally, up through the atmosphere. It carried with it the equivalent of the combustion products—in this case, most of the material from which the planetesimal was made plus seawater. The shot itself represents the ejecta, that is, the seawater and rock from the seabed that were thrown into the air. The muzzle flash leaves the gun at a higher velocity than the shot, but here our analogy breaks down because the plasma cloud would have retained much of the energy of the event. The ejecta would have fallen back into the sea and on to the coastal areas of the continents to either side, and they would have fallen fairly quickly. The plasma cloud would have condensed, and if we assume that at its center it was very dense, we may assume it would have condensed into large drops.

They, too, would have fallen quickly and close to the site of the impact.

The dust from the event would have been distributed, therefore, as a relatively high concentration around the site of the impact, falling off rapidly with distance, and much more evenly distributed over the rest of the Earth. However, while the difference in concentration between the material close to the impact site and that further away would have been marked, it is difficult to see how it could have produced the striking anomaly that we find when we compare the Danish, Spanish, and Italian clays. We need to discover some mechanism by which the dust might have been concentrated after it fell to the surface. We can imagine such a mechanism operating in the ocean, but for such concentration to occur in material that falls on dry land would have involved a process that is more complex and on the whole less convincing.

It is simplest to suppose that we are dealing with an event that occurred in the northern hemisphere, because it is in the northern hemisphere that the highest concentration of debris occurs. It is not impossible that it could have occurred in the southern hemisphere, but under normal circumstances there is little movement of tropospheric air across the equator, and we need air movement to transport the dust. We might find a way to explain such transport in this special case, but surely it would strain credulity too far to require material to be transported and then concentrated in, and (so far as we know) only in, the opposite hemisphere. Let us agree, then, that the event took place in the northern hemisphere, and let us agree also that we are dealing with an event which occurred at sea. Since the anomaly (i.e., the difference in iridium and osmium concentrations found to exist in the several sites examined) is found in Europe and nowhere else, it is reasonable to suppose that the sea in which the event occurred was the North Atlantic area.

Denmark is not in the North Atlantic, of course, and it was not in the North Atlantic 65 million years ago. So how did so much of the debris find its way to Denmark?

To answer this we must recall that the boundary clays in Denmark are famous as "fish clays." They contain numerous fossil remains of fish, far more of such fossils than are found in most places. Was this a place to which fish swam in order to die, like the legendary (but nonexistent) graveyards of elephants? Clearly this cannot be so. Fish do not swim to die in some favored hospice.

Were conditions in the sea that was then Denmark such that fish which entered it were killed, perhaps by being poisoned? Again this cannot be so, because fish are equipped with highly sensitive apparatus that makes it easy for them to stay away from poisoned water. They can be trapped by water that becomes poisoned after they have entered it, but if this were to happen once other fish would not then enter to die. There is only one feasible explanation. The fish died wherever they happened to be, most of them out in the ocean, and after death their bodies were carried to Denmark by currents and tides. This is a mechanism which operates to this day. Occasionally trawlermen will haul in a catch of dead and rotting fish, most probably remnants of earlier catches dumped overboard by other fishermen, which had been transported and concentrated in a particular place. If the ordinary movement of the waters can transport the remains of fish to a particular "cemetery," then the same mechanism can explain the transport of debris, sinking as small particles through the water, probably slowly.

What kind of a sea was it, then, that covered Denmark? We must suppose a sea into which water flows, but out of which little or no water flows. This may sound improbable, but in fact it is a fairly common circumstance. The modern Mediterranean, for example, receives seawater entering through the Strait of Gibraltar and fresh water from rivers, but within the Mediterranean itself the only loss of water is by evaporation. So water evaporates from the surface and the loss is made good from the Atlantic, but very little water, and perhaps none at all, returns from the Mediterranean to the ocean that feeds it and the exchange of water with the Red Sea through the Suez Canal is minimal. It takes eighty years for the water in the Mediterranean to renew itself—and the waters of the Black Sea, which is fed by the Mediterranean, renew themselves even more slowly.

The Mediterranean is what the United Nations Environment Program calls a "regional sea." That is to say, it is shallow, almost completely landlocked, and much at risk from pollution. What is more, it is but one of many regional seas. To understand why pollution is a particular hazard in these seas we need only look at the salinity of the water in the Mediterranean. Close to Gibraltar, where Atlantic water enters, the salinity is almost the same as it is in the open Atlantic. Further east, however, the salinity increases quite markedly, because the evaporation of fresh water from the

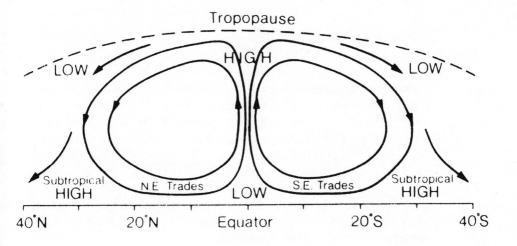

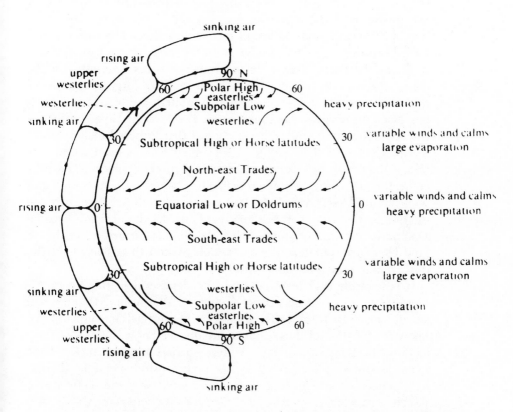

Two diagrams show the vertical and horizontal circulation of the Earth's atmosphere. Notice how, in both diagrams, very little air crosses the equator. Source: *Atmosphere and Ocean* by J. G. Harvey. Artemis 1976; Open University.

surface concentrates the salt in the water that remains. Since this is what happens to salt, it must be—and is—what happens to pollutants. They become concentrated.

Our antique Danish sea may well have been a sea of this sort. Dead fish, planetesimal pollution, anything that could not swim and so decide for itself where it wished to be was liable to be drawn into the sea and dumped there. The process need not have been rapid. There was ample time. The climate in the late Cretaceous was mild and it is more than likely that the average air temperature over the sea was as high as, or higher than, that over the Mediterranean today. Fresh water evaporated and was replaced by Atlantic water and such matter as was carried in by the Atlantic accumulated on the bottom.

In the mud on the bottom, ordinary decomposition by aerobic organisms was inhibited. We know this because of the presence of fish fossils. Had the dead fish decomposed in the usual way there would have been nothing left of them to fossilize. The water would have been deoxygenated, probably fairly stagnant, and the activity in the mud rather different. There would have been activity, for the mere absence of oxygen is no deterrent to large numbers of microbial species whose ancestors inhabited the world before it possessed an oxygen-rich atmosphere. They would have worked on the contents of the mud, releasing gases that bubbled malodorously to the surface, and in doing so they may have moved metals for which they had no use, leaving them in discrete places. The concentration of metals by microorganisms is well documented and widespread—under precisely the kind of conditions we suppose existed in the sea that eventually filled, dried out, and became Denmark. We do not suggest that the impact occurred in what is now Denmark, but that the sea which existed then scavenged the dust from a wide area of the Atlantic.

It is possible that material could be deposited in the North Atlantic from an event occurring elsewhere. Warm air rises and cool air descends to form cells, a pattern of vertical movement. We know the location of those cells today but we know nothing of their location in the remote past. Matter carried upward might have been deposited anywhere. If so, we may consider possible alternative impact sites. The first to be considered was the Tethys Sea, which was closing during the Cretaceous—between Arabia and Eurasia—and which now has disappeared entirely as continental

land masses have collided. There is not much we can say about a
Tethys crater, except that it ceased to exist millions of years ago.
The second possibility is the Tyrrhenian Sea, in the Mediterranean
basin, bounded by Italy, Sicily, Sardinia, and Corsica and, at its
center, one of the deepest parts of the Mediterranean. The rocks
surrounding the Tyrrhenian Sea bear signs of a violent origin. This
is usually taken to be the result of the intense—and continuing—
volcanic activity all around the sea, but is the volcanic history of
this cradle of Western civilization the only explanation? The
thought is intriguing, but because we can know nothing of how
material traveled to the Atlantic but can devise a means for it to
move from the Atlantic to Denmark, we prefer to place the impact
close to the mid-Atlantic ridge.

We have an explanation, then, for the much larger concentrations of metals in Denmark and for their more even distribution
throughout the rest of the world. It is time to return to the period
shortly after the impact.

The column and plume are still where we left them, although we
have removed a large amount of material from the lower atmosphere and replaced it in the sea. What of the upper atmosphere,
and inner space, though? There would have been so much material
there that the light of the Sun would have been blotted out completely. Darkness would have descended on the Earth, or at any
rate on the part of the Earth in the vicinity of the impact. For two
and a half days after Krakatoa exploded in 1883 people had to use
lamps in daytime within a radius of 80 kilometers (50 miles). For
how long, and over how large an area must this Cretaceous impact
have plunged the world into darkness?

It is difficult to know, but in considering any climatic effect it is
important to mention that the latitude of the event is of great importance. Krakatoa lies in the tropics, and dust and ash injected
into the air in the tropics has a much greater climatic effect than
would similar material released in a higher latitude. This is because
of the behavior of air masses. Air that is warmed in equatorial regions rises and its place is taken by cooler air moving in from a
higher latitude. Since cooler air occurs to either side of the equator,
it moves toward the tropics from both north and south so that two
distinct circulation patterns establish themselves, one in each hemisphere. Where the two tropical air masses meet and move against
one another, there is a small amount of mixing of the air, but in

general, material which enters the atmosphere in one hemisphere may be carried toward the equator but cannot cross it. However, there is an exception. If material enters the stratosphere close to the equator, its own energy may be sufficient for it to be carried across into the opposite hemisphere. If this happens, the material will be distributed in both hemispheres, and because it enters the tropical circulation system, which primarily is conveying air to and from the equator, it will be carried into the middle latitudes, above which another north–south circulation system operates. If the material is liable to affect climate, therefore, it will affect it globally. If the material enters the atmosphere in a higher latitude, it is likely to be contained within its own hemisphere and so affect climate within that hemisphere only. That said, we must remember we are dealing with an event far larger than the Krakatoa eruption, so what we lose in latitude we gain in amplitude!

The Alvarez team estimated that the world would have been in total darkness for up to five years. W. H. McCrea suggested it would take three to five years for all the material to have fallen— which is not the same thing. It is tempting simply to extrapolate from the Krakatoa event and assume that the darkness would last for a period of time that is a simple multiple of the Krakatoa-induced darkness derived from a comparison of the material ejected by the two events. Unfortunately, it is more complicated than that. The rate at which particles fall out from the atmosphere is exponential, and the residence time of particles in the lower atmosphere is too short for them to produce any but the most temporary effect. In our case, much depends on the way the gases mixed in the original cloud. It is a subject to which we shall return in the next chapter. In attempting to estimate the duration of the darkness, the Alvarez team was seeking to explain the observed biological effects. That is our purpose, too, and another to which we will return in later chapters. For the moment, it is enough to point out that a calamitous outcome does not require darkness to last for as long as five years or even three. One year would be ample, and several months probably would be sufficient.

Throughout this chapter we have concerned ourselves with the immediate effects of an impact from a planetesimal—not unreasonably since this is what we believe happened. All the same, how much difference would it have made had the object been a comet?

We may be making a distinction without a difference. If our

planetesimal was an Apollo body—a lump of cosmic rock with an orbit that intersects with the orbit of Earth—it may have been a "dead" comet. Most of the bodies within the solar system have orbits that can be explained by theories of the formation of the system itself. Comets and Apollo bodies are exceptions, for they have orbits that do not fit into the general pattern. This being so, it is tempting to suppose that one explanation may serve for both and that comets and Apollo bodies share a common origin, that the Apollo bodies are comets of a special kind. There is an element of hairsplitting here, as far as we are concerned, since even if they do share a common origin, the composition of Apollo bodies—solid rock—is different from the "dirty snowball" composition calculated for the more conventional comets. We need to consider the orbits of these bodies to determine how one of them came to strike the Earth—and we shall do so in more detail in Chapter 10—but for the moment it is their composition that interests us more.

It is difficult to see how the comet could have contributed enough of the metals found in the boundary clays—as we have said before. That apart, there would have been very little difference. The effect is caused, after all, by the energy expended, and as we saw earlier, a small comet moving fast may have as much energy as a large planetesimal moving slowly. The destruction of the cometary nucleus would have contributed much less solid matter, but since the mass of a body impacting at the surface forms only a small part of the total amount of material ejected into the air—which is determined by the energy of the impact—this is of little importance. Similarly, had the comet entered on a shallow trajectory, it would have burned up in the atmosphere and caused just as much heating since this too depends on the energy expended and not simply on the size of the body involved. Perhaps the chemistry of the event would have been slightly different owing to the (assumed) different chemical compositions of planetesimals and comets, but since effects are likely to have been caused by the impact rather than chemistry involving the body itself, this too matters little. In fact, a comet would have produced very much the same effect as a planetesimal, but it would not have delivered so much iridium and osmium.

It is time to abandon entirely the idea that a supernova may have been responsible. It has been suggested that a supernova explosion might have placed large amounts of particulate matter into the upper atmosphere. According to this theory, the material would not have come from the supernova itself, but would have been dust, blasted in our direction from the surface of the Moon. This would hardly matter, though, because a supernova event that occurred close enough to the Earth to blast Moon dust our way would also bathe the planet, and its inhabitants, in intense cosmic radiation. Such an event would need to be no more than about 30 parsecs (about 97.5 light-years) away from us. A supernova event of this kind could explain some of the effects that have been observed, but could it have delivered the stuff of the boundary clays —from the Moon? In fact, there is no hard evidence that there has been a supernova burst this close to us during the entire history of the solar system, and there is no evidence that any terrestrial event has been caused in this way.

We are left, yet again, with the planetesimal. We have described the force with which it would have exploded on impact, the blast of it, and the tremors that shook the entire world. We have imagined earthquakes and volcanoes, tsunamis larger by far than any that have been seen in the course of human history. In fact, we have described the force as being that of the entire global armory of nuclear weapons increased ten thousandfold and then detonated simultaneously all in one place.

The area of the impact, the North Atlantic, would have been destroyed biologically. There is no way in which any living thing could have survived. In eastern North America and western Eurasia there would have been considerable devastation, with much loss of life.

What, though, would have happened to animals living in the Pacific, in the southern hemisphere, in South America, or, for the matter of that, in western North America? Why should an incident, even one so dramatic as this, disturb the even tenor of life in Africa? In a word, why should such an impact have caused the extinction not of all species—that would not be difficult to explain, although we would not be here to explain it had it occurred—but of some?

Many of the microorganisms could survive such an event, especially those at sea. They would suffer nothing worse than a severe shaking and mixing, unless they happened to be very close to the focus of the event. They would recover quite soon. Their mysterious, anonymous lives would continue, which means that evolution itself could continue.

Evolution did a great deal better than merely continue, and if we are to understand what happened next, we must try to discover not what it was that killed so many species—we think we know that— but just how it killed them. From what cause did the species die? What shall we write on their death certificates?

7

Causes of Death

We have made much of the curious chemistry of the boundary clays that mark the end of the Cretaceous. It is time now to say a little more on the subject, for it is not only their content of metals that distinguishes them from the strata above and below them.

In the last few years a new technique has been developed that is used in the reconstruction of past climates. There are two important isotopes of oxygen (16 and 18) that exist in the Earth's atmosphere. On average, 99.76 percent of atmospheric oxygen is oxygen 16 and about 0.2 percent is oxygen 18. The proportion of one isotope to the other was determined by the original supernova that produced the elements from which the solar system is composed. It is fixed, therefore, and remains constant. When carbon is oxidized to form carbon dioxide, the proportions of the oxygen isotopes is preserved, the carbon dioxide containing them in the same proportions as they are found in atmospheric (or any other) oxygen.

Carbon dioxide is the gaseous form of a very weak acid, carbonic acid. Small amounts of carbon dioxide dissolve in water, where they react with metals to form carbonates of those metals. Some metallic carbonates, most notably calcium carbonate, are taken up by living organisms, which use them in the construction of their own bodies. The organisms most dependent on calcium carbonate are the mollusks—the snails and shellfish—and the countless microscopic species that dwell among the plankton, the mass of organisms ranging in size from single cells to the larvae and fry of many fishes which drifts near the surface of the sea and in large bodies of fresh water. However, these organisms select one oxygen isotope in preference to others, so that a biological selection mechanism exists. When the organisms die, their calcium carbonate remains, which are insoluble in water, are often preserved and they retain the selected proportions of oxygen isotopes.

The release of carbon dioxide into the atmosphere, the formation of carbonates, and the size of populations of organisms which use calcium carbonate all depend on climatic conditions, most especially on temperature. Sedimentary rocks that have been subjected to much deformation and compression over millions of years may provide little direct detailed evidence of the amount of biological activity occurring at the time they formed, but they are likely to contain the remains of organisms in the form of calcium carbonate. Consequently, an analysis of samples from those rocks that determines the proportions of the isotopes of oxygen will reveal the amount of biological activity, and from this it is possible to calculate the temperature of the water in which that degree of activity would take place. Most elements exist in several isotopic forms and oxygen is not the only one whose isotopes are selected biologically in this way.

Among all the many examinations that have been made of core samples taken from the Pacific seabed by the Deep Sea Drilling Project, some considered the isotopes of oxygen. From these examinations of samples from many different localities, we know that at the very end of the Cretaceous the temperature on the seabed and in surface waters increased by between 1 and 5 degrees C. This does not imply a general change in climate, for which no evidence exists, but it tends to confirm our picture of a large expenditure of energy at that time. The temperature increase was abrupt and short-lived.

Carbon isotopes were also measured. They showed a decrease of between 0.1 and 0.3 percent in carbon 13. Carbon has two stable isotopes—12 and 13—and a decrease in the amount of carbon 13 implies a proportional increase in the amount of carbon 12. Also as with oxygen, biological organisms select one isotope of carbon in preference to the other. In other words, a decrease in the amount of carbon 13, far from implying a reduction in the amount of carbon as a whole, can imply an increase in total carbon, provided the carbon is of biological origin. Commenting on this in 1980, Kenneth Hsü (*Nature*, Vol. 285, p. 201) has said it is "equivalent to a change produced if the whole of the terrestrial biosphere were put into the ocean." Of course, the terrestrial biosphere was not "put into the ocean"—organisms that died on land remained on land—

but a very large number of marine organisms, containing carbon 12 preferentially, did die.

Is there an alternative explanation to the obvious biological one? Since the proportion of carbon 12 to 13 is constant for the carbon that is present on Earth—another product of the stellar conditions in which heavy elements were formed—if we are to have an alternative explanation, it would seem that we need a source of lighter carbon—carbon much richer in carbon 12. Hsü suggests that this lighter carbon might have been contributed by the body itself. While the carbon found in meteorites and in chondrites (the lumps of carbonaceous material found in some stony meteorites) has much the same isotopic composition as terrestrial carbon, and while the carbonate they contain consists of very heavy (carbon-13 rich) carbon, Hsü proposes that a comet might contain carbon monoxide and carbon dioxide which, because of the manner of their formation, were markedly lighter. His argument is speculative and not particularly compelling. He proposes that an impacting body might be capable of delivering the requisite quantity of light carbon, but in this case we have too much light carbon, for if we allow that the event caused the death of large numbers of organisms, the light carbon contained in their remains must have gone somewhere and the most likely place to find it is in the sediments deposited at the time. All in all, we would prefer to regard the reduction in carbon 13 as a measure of the extent of the biological effect of the event.

We are looking for ways in which we may reconstruct the events that led to the deaths of organisms. The task may seem superfluous if we have convinced you that such events themselves were caused by an event of the magnitude we have tried to describe. Our forensic minds cannot be satisfied with such immediate consequences, however. Just as no doctor today can accept "old age" as a cause of death, we cannot accept "struck by planetesimal" unless we can discover how it was the planetesimal achieved its dire effect.

Before proceeding further, perhaps it will be as well if we consider what that effect was. What was made extinct and what survived? If we can do that we may obtain clues, and it may be possible to construct theories, albeit tentative ones, about changes in the environment that will injure some organisms but not others.

About 70 percent of all species living at the end of the Cretaceous Period are absent from the Lower Tertiary. That is the extent of the extinctions as a whole, but how were the extinctions distributed?

In the seas, the calcareous plankton—the plankton containing and dependent upon calcium carbonate—suffered severely. The foraminifera, for one, were almost made extinct, and while they recovered again later, they have never since been as abundant as they were in the Cretaceous seas. Foraminifera are protozoans, simple animals many of which are single-celled and microscopic, but some of which are multicelled and have grown to 15 centimeters or so in diameter. Those giants among foraminiferans are platelike in shape and are found in limestone, in Syria, among other places. All members of the group live within tests, or shells, and when they die their tests sink to the seabed, where they used to accumulate. The famous chalk cliffs of Dover, in southern England, are made from the remains of foraminifera, and a thick stratum of them at an altitude of 6,700 meters on Mount Everest shows that what today is high mountain was once seabed.

Throughout history there have been some 20,000 foraminiferan species, only about 1,200 of which are alive today, most of them marine. A characteristic all of them share and have shared in the past is a high degree of adaptability to their environment, often at quite a local level, and their great sensitivity to changes in that environment. Some live on the seabed. Others float with the plankton near the surface of the sea. The ocean sediments rich in their remains are called "globigerina ooze." Some foraminifera are especially sensitive to changes in temperature and are either present or absent, and—more intriguingly—coil their shells to the left or to the right—at different temperatures and so can be used as indicators of past climates.

The nanoplankton was also very severely affected. This diverse collection of organisms is grouped together only on the basis of the size of individuals and the region in which they are found. The prefix "nan" means "dwarf"—and nanoplankton are very small organisms indeed that live among the plankton.

Technically, nanoplankton are plants and manufacture sugars by photosynthesis; others are animals, so called because they feed upon the plants or upon one another. The planktonic plants—the phytoplankton—are the principal ecological producers in the open sea, where they assume the role played by green plants on land. Among the phytoplankton, it was assumed until fairly recently, the most important producers were the diatoms—unicellular organisms enclosed within a transparent shell of silica, the largest of them barely visible to the naked eye. Although many diatoms live singly some join together in chains and others, while living singly, have shells that are drawn out into long hairs. These characteristics make them the most obvious members of the phytoplankton. More recently, however, it has been found that the most important producers are still smaller organisms, the flagellates. They used to be classified as plants, although their mode of nutrition places them somewhere between plants and animals, and today a new classification system is used. They possess flagella by means of which they swim and to create small currents that waft particles of food toward them. Flagella (the singular is "flagellum") are tiny projections from the cell walls, looking for all the world like tails or fine hairs, depending on how many of them an organism possesses. They are not confined to free-living single-celled organisms. Many of the cells in our own bodies possess them, and it is by the movement of flagella that our senses transmit information about our surroundings to the nerves which carry it to our brains. Spirochetes are single-celled organisms that consist of little but a single flagellum, and some biologists suspect that all flagellates consist of a cell with a captured spirochete. The suspicion is given force by the fact that spirochetes and flagella are constructed in precisely the same way no matter where they may be found.

Since these members of the plankton are the most important producers of food in the seas, compared with which all the larger animals are consumers, any change in the size of their populations that increased or decreased the level of productivity in the seas necessarily would be reflected in the populations of larger animals.

Perhaps we should digress for a moment and explain the way in which energy, and so food, passes through oceanic or any other food webs.

The process is cyclical and we might begin a description of it at any point in the cycle, but the conventional place to start is with the autotrophs, the producers. These consist principally of green plants—of every size from single-celled algae to the largest trees of the forests, and in the world as a whole it is the algae that are the more important in terms of the volume they produce. By photosynthesis they convert water and atmospheric carbon dioxide into sugars, sunlight supplying the energy that is used for the process. At the same time, microorganisms living mainly in the soil or in seawater are able to "fix" simple mineral substances, using them in their own bodies, and, on death or as waste products, supplying them as compounds that can be taken up by plants. Nitrogen, for example, although abundant, is useless to plants in the gaseous form, and plants obtain it as nitrate after it has been processed by certain groups of bacteria. The autotrophs are eaten by other organisms, which are, by crude definition, animals. More accurately, they are called "heterotrophs," organisms that cannot manufacture food but must obtain it in a ready-made form and then modify it metabolically in accordance with their own requirements. They are consumers. If they subsist wholly or mainly on a diet of producers, they are primary consumers, that is, herbivores. Carnivores feed on herbivores and so they are secondary consumers, and if they feed on other carnivores they are tertiary consumers.

In this half of the cycle, solar light and heat are used to supply the energy for chemical reactions by which simple inorganic molecules are used to construct more complex molecules which are then assembled into the forms of the living organisms that we see about us and that sustain those organisms.

The other half of the cycle is concerned with the breaking down of those large molecules into the original simpler ones. These are returned to the sea, to the water in the soil, or to the air, so that the process of construction may begin all over again and the cycle continue. This second part of the cycle, decomposition, is mediated by a hierarchy of organisms that parallels the hierarchy of composition, but it is more heavily reliant on the activities of microorganisms.

We may imagine, then, species linked by their dietary relationships into a complex network. There is another way in which we can picture the relationships involved in the cycle, based on the observation that organisms can be arranged hierarchically. If we consider the constructive part of the cycle and measure the total mass—the biomass—of all the organisms engaged in it at each level —producers, primary, secondary, and tertiary consumers—we can represent this diagrammatically, with a block for each level, the height of all blocks being the same but the width of each representing its biomass. It is usual to draw the diagram with the widest block at the bottom. When we do this, we find it forms a stepped pyramid shape, called the "biotic pyramid," whose significance is clear. We have placed the producers at the bottom and the primary, secondary, and tertiary consumers above them. Obviously, a consumer must not eat so much that the species on which it depends for food is unable to reproduce and so replenish the larder, and in general the biomass at one level is approximately one tenth of the biomass at the next level below.

Anything that alters the biomass at any level will affect the pyramid as a whole, but the extent of that effect depends on where the alteration occurs. At most levels, an increase in biomass will cause the food resource from below to be stretched too thinly; at the same time, it will increase the amount of food available to the organisms in the level above, so that the population will be reduced and the balance restored. Consider, for example, the effect of a sudden increase in numbers among the herbivores. The enlarged population will place an increased strain on the food supply and eventually food shortage may lead to migration or even starvation, this restoring the population to its former size. Before this happens, however, it is possible that those carnivores which prey upon the herbivores in question will undergo an increase in their own numbers. Predation will increase and, again, the effect will be to restore the balance which existed formerly. The example is highly theoretical, of course: in the real world, relationships are much more subtle. If the bottom level is increased, there will be more primary consumption to restore the balance, but if the bottom level is decreased, the repercussions will be felt throughout the entire pyramid. The end result is likely to be a smaller pyramid, possibly with the top level—of tertiary consumers—removed entirely since there is no longer sufficient food to sustain it.

Clearly, then, a severe reduction in the phytoplankton may well lead to the extinction of species at the top of the marine pyramid.

It is the sudden and drastic reduction in the foraminifera and nanoplankton that marks the end of the Cretaceous, and it is described clearly by J. Smit and J. Hertogen (*Nature*, Vol. 285, p. 198, 1980): "In the more than 200 m thick youngest Maastrichtian marls of the Caravaca (Spain) section the rich, tropical associations of planktonic foraminifera and nanofossils shows no significant changes up to the very last centimeter; here almost the entire association disappears within 0–5 mm. . . . this implies that the extinction took place within about 200 years." In our view, the extinction cannot have taken longer than two hundred years; it may well have occurred more rapidly.

What is especially interesting is that the species of nanoplankton that survived the event are known to tolerate or even to prefer conditions that most species cannot tolerate.

Ammonites disappeared finally, in the impact event, and on land all animals weighing more than about 25 kilograms were killed. Plants on land were not much affected. As we saw earlier (Chapter 3), the modern angiosperms were well established by the end of the Cretaceous, and they continued to evolve in the Tertiary. There is no reason to suppose the event had any effect at all on their evolutionary development. Small freshwater animals also survived relatively unscathed.

The event, then, produced patchy effects. Marine plankton was affected very severely, the ammonites disappeared, large land animals disappeared, but land plants and small freshwater animals survived. What could produce this kind of pattern?

Let us consider first the idea that the amount of dust ejected into the atmosphere was sufficient to blanket out sunlight. Walter Alvarez proposed that idea, but it was criticized on the ground that no matter how large the mass of particles that enters the atmosphere, the time taken for them to fall out remains fairly constant. This is because—so runs the argument—the denser the cloud of particles, the greater is the chance of collisions between particles. When particles collide, some of them will adhere to one another. This increases their weight and so they fall, and their "residence time" in the atmosphere is inversely proportional to the number of them that are present. Quite simply, the more of them there are, the sooner they will fall, and since any cloud of particles in the lower atmosphere falls out at an exponential rate, the time taken for a cloud to disappear is much the same regardless of its initial density.

What really matters, therefore, is not the fate of particles in the lower atmosphere but of those in the stratosphere. Particles in the lower atmosphere cannot remain suspended for long enough to produce more than the most transitory effects.

In the case we are considering, the cloud was formed mainly by the condensation of a cloud of gas, not by the ejection of solids into the atmosphere from below, and much depended on the density of that cloud. Fine particles would form only if the cloud were tenuous. If it were dense, it would form large drops, which would fall like rain, at once. Probably most of the precipitation would occur within a particular area.

The criticism is open to two serious objections. In the first place, let us suppose that many very small particles did form. Do small particles in the air actually collide? Experiments conducted many years ago into the dispersal of aerosols from an atomizer suggest they do not and that even in a very dense cloud of particles, collisions are not much more common than they are among the densely clustered stars at the center of a galaxy. Collisions do occur, but not often, and the residence time of such particles is not affected significantly by them.

A second objection concerns the actual density of cloud that would be needed to block out sunlight, what is known scientifically as the "column density." Consider the leaf canopy in a broadleaf forest. If you stand on the forest floor and look upward, it may be quite impossible to see the sky. All you see are leaves. This is not because the leaves are packed together closely at one level, but because they overlap with one another at many levels. What matters is not only the number of leaves (or particles) at a particular altitude, but the number that are interposed between the surface and the top of the atmosphere. Because this particular cloud was ejected with wholly unprecedented amounts of energy, we may suppose it to have been distributed vertically from the lowest to the highest level and into near space itself. If this was so, the cloud may have been very tenuous indeed, as measured at any particular level, and yet have succeeded in blocking sunlight very effectively.

We can see this effect demonstrated most effectively on Venus. As viewed from space, Venus appears shrouded in dense cloud. It is impossible to catch even a glimpse of the surface. When instruments were soft-landed on to the surface of the planet, however, and pictures were received from them, they showed dry, rocky ground and a sky that was blanketed not in cloud, but in a fine haze. Horizontal visibility within the venusian (an ugly word: the correct adjective is "venereal," but it is little used by space scientists) atmosphere is good. The thick cloud in fact is a column of thin haze that is very deep indeed. Nor is this all. If some of the ejected particulate matter is in orbit, it is likely to fall back into the atmosphere gradually so that the atmospheric stock is replenished from above.

We know a good deal about the behavior of particles in the atmosphere. In 1963 Mount Agung, a volcano on Bali, in Indonesia, erupted and its effects were measured in detail. After the eruption the temperature in the stratosphere close to the equator rose by 6 to 7 degrees C, and it remained at 2 to 3 degrees C above the preeruption level for several years. No change in the temperature of the troposphere was recorded. In that case, however, all the ejected material remained in the atmosphere. If material leaves the atmosphere to enter a low orbit about the Earth from which it returns slowly, we cannot use our knowledge of the "Agung effect" or, indeed, such records as there are of the earlier Krakatoa eruption.

It is most unlikely that material ejected into the atmosphere could cause prolonged darkness, but material entering continually from above the atmosphere might possibly do so. It is conceivable that the ·Alvarez team may be correct—that the world was plunged into darkness and remained in darkness for some time; the Alvarez team suggests for five years. Would such darkness have produced the effects we know occurred?

Photosynthesis would have ceased if the darkness was total, and would have been much reduced if it was partial. Many green plants would have died, and the smallest and simplest of them would have died first. The marine phytoplankton would have gone quickly, and with its disappearance the zooplankton which feeds on the phytoplankton would have been reduced or eliminated. The darkness would account quite satisfactorily for the marine extinctions.

Why were the land plants affected less severely? To answer this we should begin by asking what we mean by "darkness." This may sound obvious, but as far as plants are concerned, "darkness" consists in the absence of sunlight, and sunlight may be absent as far as plants are concerned but not as far as animals are concerned. It is possible for ash to coat leaves, and especially if the leaves are wet when the ash falls, it may adhere to them firmly. No more than a thin, but total, covering of ash may be sufficient to inhibit photosynthesis. In effect this is the same thing as "darkness."

Having established that, a possible answer to our question is not too difficult to find. The plants themselves may well have died, but their seeds, sealed securely within watertight (in some cases, partly fireproof) cases and hidden below the surface of the soil, would have remained viable for some years and perhaps for many years. Cereal grains, for example, have germinated after centuries spent in the tombs of former human dignitaries; they will do so provided they are kept dry and cool. It is entirely possible, therefore, for most or even all of the surface plants to have died and years later to have reappeared.

Such darkness may also explain the apparently more severe effect on the more primitive plants. Most of these would have been ever-green and not adapted to a life cycle that involved the regular loss of leaves. When photosynthesis ceased many of them would have died. Deciduous plants, however, would have shed their leaves and grown new ones.

We are not saying that the deposition of dust did produce such an effect—only that it may have. The Alvarez team may still be right, and the world may have become dark as well, or both effects may have operated together.

Why did this loss of vegetation kill only the large animals? The first part of the answer to that is that we have no reason to suppose that only large animals were killed. All we know is that the large animals became extinct. Small animals may have died in large numbers, and probably did, but a sufficient number remained for populations to expand again when the plants returned.

The small animals have two advantages over large ones in such adverse circumstances. They are better able to adjust their behavior to constraints imposed by the environment and they require less food than large animals.

A period of prolonged darkness would have been similar to a long winter night. Many small mammals are largely nocturnal in their habits and so the darkness would have been a benefit to them rather than a hindrance. Wood mice, for example, are deterred from foraging even by bright moonlight. In winter, small mammals spend most of their time in nests that are well lined with leaves or grass. Insulated against the cold and inactive for hours or days on end, they need much less food than they do in summer. Many of them store food for the winter, but not very efficiently, and a lack of provisions is not necessarily serious. When conditions permit and animals are hungry, they forage.

Small mammal populations may be dense at the commencement of the winter. (In the case we are imagining, there would have been no advance warning—perhaps there would have been no autumn—but, even so, dense populations may well have existed.) They are dense because the autumn and winter follow the main breeding season, so young are surviving and most of the older individuals have died earlier in the year, and because transient individuals, which spend the spring and summer wandering about the countryside, settle down and remain in one place for the winter and so are counted into the overwintering population.

As the winter progresses, especially if it is harsh, mortality may be high. But mortality is selective—the first to die are the old, the sick, and the small, weak individuals. They are bullied by the large individuals, who are able to obtain more than their fair share of the food that is available. However, the next group to suffer consists of these large individuals. Aggressive they may be, but their large body size means they need relatively large amounts of food, and if they cannot obtain it, they die. The winners in this contest are the young, healthy, medium-sized animals whose food requirement is modest. Studies of overwintering small animals have shown that although mortality from starvation may be high, the individuals that survive are hardly ever malnourished.

In our imagined event, we are supposing a very long winter, of course, but this need not prove too troublesome. The plants would have died down, but small mammals feed on seeds, on insects, and their larvae and pupae, and on other small, invertebrate soil animals; the soil invertebrates themselves feed mainly on plant and animal wastes and on one another; and there is no reason to suppose that these foods were anything but plentiful.

Small animals have shorter life spans and breed more frequently than large animals. This, too, would be an advantage under the circumstances we are supposing, for colonies of small animals would have tended to respond to the darkness as though it were merely a prolonged night or a winter. They would have remained in their colonies where encounters with members of the opposite sex would have been frequent. The darkness might well have disturbed those hormonal cycles that are related to changes in day length, but females would have continued to have come into estrus—irregularly perhaps—and breeding would have been sustained. The large animals, on the other hand, would have started with a much lower population density since each individual would have required a much larger land area to supply its food. In the darkness the response might well have been to disperse in search of food—herd animals keep in contact with one another by sight and to a lesser extent by sound, but we are familiar only with herds of mammals: herds of reptiles are strange to us and we have no way of knowing how they would have behaved in the herd. If they did disperse, encounters between individuals would have been reduced, and therefore matings would have become less frequent. Among animals which reproduce only once a year—or less—this might lead quickly to a declining population. If they did not disperse and the integrity of their herds was maintained, their response to a reduction in food supply, assuming they responded at all, could only have been to increase the frequency and extent of their migrations or to begin migrating for the first time. In darkness, or near darkness, contact among herd members would have been much more difficult to maintain, individuals would have been separated and there might well have been an increase in conflict among wandering herds trespassing on the ranges of one another. Since such large surface animals must have fed mainly on growing vegetation rather than on seeds, the pressure on them must have been far more acute.

In the long night that might have ended the Cretaceous, therefore, we may picture the small animals continuing their lives in a slightly different way, but with no great inconvenience, while the large animals suffered greatly.

The small animals that survived in fresh water might have done so partly by hiding in the bottom mud. There are desert fish which can survive for years in this way until the rare desert rain turns the arid landscape to meadow and their mudholes to ponds, in which they may meet one another and mate. The microbiological activity in the muds is concerned with the decomposition of organic detritus—of which there can have been no shortage—and it does not require sunlight. Indeed, it may be inhibited by sunlight. To the extent that small freshwater animals depended on such activity, they would have suffered no inconvenience.

It is not very difficult to find explanations for the deaths of large animals and the survival of small animals, but in doing so we have overlooked what may be one problem. Some modern reptiles are inconveniently large. The crocodilians, some of the large constrictor snakes such as the pythons, monitor lizards such as the Komodo dragon, some iguanid lizards including the common iguana, the leathery turtle and the green turtle—all are reptiles and all of them weigh more than—in some cases much more than—the 25 kilo-

On May 18, 1980, Mount St. Helens erupted in a massive display of volcanic power, sending an enormous plume of volcanic ash and dust into the stratosphere. (*J. Stewart Lowther*)

Top. The devastation caused by the Mount St. Helens activity is seen in this "moonscape" photographed over two years after the initial eruption. The volcano itself is seen in the background, its top completely blown away. (*Dr. R. Spicer/Science Photo Library*) **Below.** Trees near Mount St. Helens killed by a combination of heat and ash from the eruption. (*Dr. R. Spicer/Science Photo Library*)

grams that seemed to mark the threshold for survival into the Tertiary Period. Various explanations for this curious anomaly have been proposed. The turtles, being marine species, perhaps were protected by being far out at sea, out of harm's way. Today these turtles are confined to the southern hemisphere; if they were so confined at the end of the Cretaceous, that fact may provide explanation enough for their survival—they were out of range. On the other hand, is it possible that those which lived in the northern hemisphere were killed and that the southern hemisphere survivors have never since migrated across the equator? Perhaps the crocodilians, which live today in both hemispheres, were half covered in water, with eggs carefully protected in well-made nests. That is not very convincing —did other reptiles not protect their eggs? Snakes appeared first during the Cretaceous, but it may be that modern large snakes evolved later than the Cretaceous from smaller ancestors. Snakes are the most recent reptiles to have appeared. The iguanids appeared after the Cretaceous, but two specimens have been found from the European Jurassic that may be ancestral to them. The monitors present a more difficult problem, for they belong to an ancient family that very certainly was present during the Cretaceous.

The evidence is far from conclusive, but it is possible at least that we must find a way to explain the survival of a few reptiles whose body size would have exceeded the threshold. Has the threshold been calculated incorrectly?

We know that the mammals survived and that the mammals living at the time included some about the size of large domestic cats (probably they weighed about 3 kilograms). Being mammals, they were warm-blooded—the scientific term is "homoiothermic" (same-heated). Reptiles are cold-blooded—the scientific term is "poikilothermic" (variously heated). But the English words "warm-blooded" and "cold-blooded" are misleading. An animal—any animal—functions efficiently only between certain temperatures, although the optimum operating temperature varies from one species to another. If the temperature falls too low, the animal becomes sluggish in its movements, its heartbeat slows, and its rate of respiration falls. In this state it may survive, but it can neither feed nor reproduce. If its temperature rises too high, it may die quickly as its internal chemistry is disrupted. An active reptile has a body temperature that is much the same as that of a mammal—indeed, it may be a little higher—and in winter, hibernating mammals allow their body temperatures to fall to within a degree or two of the ambient air temperature. So homoiotherms can be cold, poikilotherms can be warm, and the effective temperature at which an animal of either sort can be active is the same. It is true, however, that most poikilothermic animals can survive in temperatures that are only a degree or two above freezing—there are fish that are able to remain active in arctic waters actually below freezing point. It is easier to survive low temperatures than high ones.

The difference between the two types of animals concerns not their body temperatures but the means by which those temperatures are regulated. A poikilothermic animal warms itself by basking in the sun, cools itself by seeking shade or water, and is sufficiently sensitive to the warming or cooling of its own body to be able to tell in advance when it will need to move from its position to one that is warmer or cooler. A homoiothermic animal possesses internal sensors and physiological mechanisms—such as sweating, panting, and shivering—by which it may warm or cool its body without recourse to anything outside itself. This enables a homoiotherm to inhabit regions that are beyond the range of most poikilotherms because they are too warm or too cold. (There are exceptions to the rule—poikilothermic amphibians such as frogs thrive in high latitudes, adders are found inside the Arctic Circle, and in hot deserts poikilothermic reptiles and invertebrates fare as well as mammals.) The real advantage, and the one that enables adders to live where rival snakes cannot, is the ability to give birth to live young, which increases greatly the chances of survival for the young and reduces the investment animals must make in reproduction. We shall return to this in Chapter 9, where we discuss the evolutionary implications of the event.

At the same time, being homoiothermic carries a penalty. Body heat in homoiotherms is generated by the digestion of food and by muscular activity, such as shivering; the removal of heat may also involve activity, such as panting. Thus the regulation of body temperature by internal means requires the expenditure of energy and this increases the amount of food the animal needs. Deprived of food, a shrew, for example, will starve to death in a matter of a few hours. It must eat almost constantly. It does not need all this food in order to repair its body tissues nor to provide the energy it uses in respiration or movement—it needs it in order to generate body heat. Its small size means that its mass is small in relation to its surface area and it loses body heat rapidly. Large homoiotherms suffer from the opposite handicap. Their relatively small surface area makes it more difficult for them to lose body heat and to remain cool.

The fact is that for homoiothermic animals the food penalty is very large indeed. It is quite constant from species to species. Samuel Brody (*Bioenergetics and Growth*, New York: Reinhold, 1945) calculated that to maintain their basal metabolisms, all homoiothermic animals expend energy equal to their body weight in kilograms raised to the power of 0.73 multiplied by 70.5 calories. This means that a warm-blooded animal needs about 10 times more food than a cold-blooded animal of similar size.

Let us now "convert" our homoiothermic Cretaceous mammals into poikilothermic animals with a similar food requirement. To do this, we simply multiply their body weight by 10 and so produce animals with a "poikilothermic equivalent" weight of about 30 kilograms. The small mammals would thus have been at, or just a little above, the extinction threshold if we assume the principal threat to life to have been starvation. They survived; perhaps some reptiles weighing a little more than 30 kilograms survived, but other, larger, reptiles did not.

Most of the dinosaurs of the late Cretaceous were large. The tendency to evolve into large forms seems to have been quite general among them. There were exceptions, however, and some oddities. *Saurornithoides*, for example, was a bipedal carnivore that moved swiftly and hunted by night. True, as an adult it grew to a length (including the long tail, long neck, and long slender limbs) of about 2 meters and weighed more than 25 kilograms, but it seems to have hunted mammals, and to have hunted them by night. Why could it not survive? The answer may be found in studies of modern reptiles, whose young receive from their parents little or no protection and no training; they enter the world fully mobile, with their senses fully developed, and well able to fend for themselves. The high mortality among them is due to predation from larger animals. Is it not likely that among the smaller carnivorous dinosaurs, whose young would have weighed less than the critical 25 kilograms at least until they are close to reaching sexual maturity, there would have been a few surviving species? All that would have been required was the survival of the young just long enough for each generation to reproduce itself. After that the large adults might have died. Natural selection might have favored the smaller individual adults who reproduced more prolifically, and those juveniles who matured at an early age and when they were small. There should have been survivors.

How did some of the reptiles adapt to a nocturnal life style? As poikilotherms, how did they manage to keep warm enough to be active? Most modern reptiles are diurnal. The only nocturnal species are not specialized for feeding by night and do so only when the weather is warm. We assume that the Cretaceous reptiles were poikilothermic, but why? The main reason is that modern reptiles are poikilothermic, but it does not follow that all reptiles have been so throughout the whole of reptilian history. We also tend to assume that natural selection must favor homoiotherms, but as we saw earlier, and as we shall suggest again in Chapter 9, homoiothermic animals incur penalties as well as advantages.

If we were to suppose that some of the Cretaceous reptiles were homoiothermic, their disappearance is much easier to understand, for we must multiply by 10 the amount of food each of them needed. When we do so, we can see that not even the young of even the smallest of such dinosaur species could have survived, while the very large animals could not have survived whether they were homoiothermic or not, for their food demand, even as poikilotherms, would have been far in excess of what was available. All would have starved. However, those reptiles that were poikilothermic would have fared much better, and perhaps that is why all modern reptiles are cold-blooded.

There is another reason for supposing that all large animals which dwell on land are likely to possess some internal means for regulating their body temperature: external means probably are inadequate. An animal's ordinary metabolic activity generates heat, but it may lose heat at the body surface. When the body surface is small in relation to the volume of the body in which the heat is produced, heat retention can prove dangerous. No large mammal is especially tolerant of heat and those that live in the tropics wallow in water or mud in order to cool themselves. Yet mammals possess internal regulatory systems. An animal that lacks such systems will find hot weather much more burdensome. It is believed that some of the very large extinct reptiles were partly aquatic, but there were many more, and especially late in the Cretaceous, that were not. The large carnosaurs hunted on land. They were committed to bipedalism—two-leggedness—with large, strong hind limbs and much reduced, weak forelimbs. They could not possibly have moved on four legs, even temporarily, and there is no way they could have moved into or from water deep enough to immerse the upper parts of their bodies. Some of the hadrosaurs—the duck-billed reptiles that lived in herds and were numerous in the late Cretaceous—had webbed digits on their forelimbs, but if they once lived in aquatic habitats, it is apparent that they had ceased to do so long before they became extinct. They had hooves on their hind limbs, and the stomach contents of fossil specimens show them to have been feeding on fruits, twigs, and pine needles, a diet to which their teeth were well suited and which is typical of dry-land habitat. They would have found it difficult to keep cool and it is reasonable to suppose that they, and many other large reptiles, in fact were homoiothermic.

If the idea of homoiothermic reptiles still seems outrageous, perhaps we should recall that so far as we know both birds and mammals have reptilian ancestors. Certainly, birdlike and mammallike reptiles existed. Were these reptiles homoiothermic or was the ability to regulate body temperature internally acquired after the true birds and mammals appeared? If this is so, we must suppose that at one time there were poikilothermic birds and mammals—an idea that is no less outrageous! We cannot have it both ways. Either some reptiles were homoiothermic or some birds and mammals were poikilothermic.

It is interesting to note here that the ability to regulate body temperature by internal means seems to have evolved slowly, as part of the higher nervous system, and it can be lost or overridden. We have mentioned hibernating animals, which allow their body temperatures to fall in winter because food is absent, the air temperature is low, and they are unable to maintain a high temperature. Other homoiothermic animals can behave as though they were poikilothermic under anesthesia, which seems to impair the heat-regulating mechanism; the mechanism can also be impaired by injuries to the spinal cord, which can turn even a human into something close to a true poikilotherm. Young birds are unable to control their body temperatures; it is an ability they develop as they grow.

We see, then, that a prolonged period of darkness caused by particulate matter ejected by the planetesimal impact and by volcanic activity following the impact is feasible. We see also that such a period could produce the pattern of extinctions that occurred. Surely, though, darkness that persisted for years would have led to a major climatic cooling. In the years following the 1883 Krakatoa eruption the weather generally was markedly cooler and this is not the only example available to us. The winter of 1784 was unusually harsh and Benjamin Franklin attributed this, probably correctly, to what he called a "dry fog," caused by dust from a volcanic eruption. It is not clear what Franklin meant by his "dry fog," but we do know that any kind of particulate haze in the troposphere would have led to a warming, so the "fog" must have been in the stratosphere (see p. 118) where perhaps he observed it as something like a thin and persistent layer of high cloud.

We may owe the story of Frankenstein and his monster to a volcanic eruption, for quite possibly it is this that produced the bad weather of the summer of 1816, and it was that poor weather which kept Percy Bysshe Shelley, Mary Godwin, Claire Clairmont, and Lord Byron indoors during their visit to Geneva and provided the romantic circumstances that led Mary to write her novel *Frankenstein*, Byron to write part of his *Childe Harold*, Claire to conceive Byron's child, and Shelley and Mary to plan their marriage, which took place later in the same year. Bad weather is not always unproductive! Byron was in Geneva in the first place because the scandal over his relationship with his half-sister had provoked his wife into leaving him, taking with her their baby daughter. The daughter, Augusta Ada, grew up, married, and as Ada Lovelace became an assistant to the English mathematician Charles Babbage and the world's first computer programmer.

We have evidence for a warming of the seas as an immediate result of the planetesimal impact, and for a general but gradual cooling during the latter part of the Cretaceous, but no evidence at all for a rather sudden cooling on a global scale. Quite certainly there was no glaciation, and a major climatic deterioration is almost bound to be marked by glaciation, at least in very high latitudes. Does this not rule out the darkness theory?

If the world were in total darkness for a period of years, it does not follow that the surface would necessarily cool. In our case, we know that much dust must have entered the atmosphere. We know, too, that there was no major cooling. We have to conclude, therefore, that we are dealing with an event that is quite different from the Krakatoa, or any other, volcanic eruption. We deceive ourselves if we assume the event to have been much like Krakatoa, merely extrapolating the effects to take account of the magnitude, find no evidence for the effects we take to be inevitable, and so dismiss the event itself.

Minute particles of solid matter in the atmosphere can affect climates, but the ways in which they do so are complex and depend on the particles themselves and their location. To understand this we must look a little more closely at the atmosphere itself. This is what we shall do in the next chapter, where we consider ways in which an impacting planetesimal might alter temporarily the composition of the air itself, perhaps adding to its opacity and so contributing to the darkness, but most certainly producing conditions inimical either to plant life, or to animal life, or to both.

So far as the prolonged darkness itself is concerned, for the moment we will content ourselves with saying that it is possible for the world to be dark yet to remain warm. Such a darkness can be adduced to account for the extinctions, both in extent and in distribution, and there is no reason to suppose it did not occur.

8

Air Pollution on the Heroic Scale

After Mount St. Helens, in Washington State, erupted so spectacularly on May 18, 1980, burying 7 million hectares of land beneath ash, many concerned but witty citizens complained to the U. S. Environmental Protection Agency (EPA). In places the ash was 75 centimeters (30 inches) deep and coated the leaves of plants up to 400 kilometers (250 miles) away thickly enough to prevent photosynthesis almost completely. The citizens wanted to know what regulations the EPA was planning that would make it illegal for volcanoes to erupt in this messy way in the future. After all, had the mountain been a factory, no character witnesses attesting to its (former) decorative value or its (present) popular entertainment or scientific value would have protected the owner from a long jail sentence and civil actions that would have cost him his entire profit for the next 10 million years. The good citizens were joking, of course, but like all good jokes theirs had a kernel of profundity.

We worry about the pollution we cause, and rightly so, but we should not forget that when she turns her mind to it, Mother Nature can and does pollute on the grand scale, and her pollution is no less damaging than ours.

After the Mount St. Helens dust had settled, the Washington air was as pure (or impure) as it had been before the eruption. The ash that had been in the lower air—the air people breathe—was on the ground. Immediately after the eruption, breathing was difficult, but within a very short time it ceased to be so. The point is that the dust settled quickly. At the same time, the gases in the plume, which rose high into the air, reacted and left behind a cloud of very fine droplets of sulfuric acid which drifted across the world. A team of European scientists (M. Ackerman, C. Lippens, and M. Lechevallier) took photographs by means of remote control cameras carried by balloons to altitudes of about 35 kilometers (150,000 feet) and showed that while the stratospheric air over Europe was reasonably clear on May 7, by June 5 a white haze of the droplets, varying from 100 meters (330 feet) to several kilometers in thickness had formed a blanket through which the cloud tops below were barely visible. They and many other people expected the haze to reduce the amount of incoming solar radiation and so to cool the air below, producing a poor summer and a harsh winter. As it turned out, they were wrong. The winter of 1980–81 was rather mild.

The example illustrates the danger of assuming that in meteorology, the obvious will happen. Nothing that happens in the atmosphere is simple and much of it is understood poorly or not at all. An impacting planetesimal would place much larger quantities of solid particles, as well as haze, into the air than any single volcano, but it does not follow that the weather would become colder as a result.

We should consider the atmosphere as being divided into two layers, as an inner shell and an outer shell separated by a distinct boundary. (In fact there are more outer layers, but they need not concern us now.) The inner shell, or lower layer, is called the "troposphere" and it extends from the surface of the Earth up to the boundary, called the "tropopause," whose altitude varies from place to place and from day to day but which is sometimes as low as 10 kilometers (about 32,000 feet) above the poles and sometimes higher than 20 kilometers (about 65,000 feet) above the equator. Many modern military and civil aircraft fly regularly above the tropopause in the stratosphere, which is the outer shell, or upper layer. It extends to a height of about 80 kilometers (260,000 feet) where another boundary, the stratopause, separates it from the layers beyond.

Both troposphere and stratosphere are composed of air: roughly 78 percent nitrogen, 21 percent oxygen, and the balance carbon dioxide, argon, and other gases, with a variable amount of water vapor, almost all of the water being confined to the troposphere. In the middle layers of the stratosphere, mainly between about 15 and 30 kilometers (9 and 18 miles) above the ground, incoming ultraviolet radiation causes the dissociation of ordinary molecular oxygen into its constituent atoms and some of these atoms recombine in groups of three instead of two, so that oxygen becomes ozone. This is the ozone layer.

An important difference between the troposphere and stratosphere occurs because of the temperature of the air in them. In the troposphere the air temperature decreases with increasing height. The extent of the rate of decrease varies from place to place and from time to time, according to the weather conditions. Local temperature inversions may produce temporarily a situation in which warm air lies trapped above cooler air so that temperature actually increases with height, but the usual pattern is one of falling temperature with increasing altitude. At the tropopause the temperature is between —55 and —90 degrees C. In the lower regions of the stratosphere this temperature remains constant, but at higher levels it begins to rise again, reaching a peak of about 2 degrees C at a height of about 50 kilometers (30 miles), then falling again in the upper stratosphere until, at 80 kilometers (50 miles), it is at about —83 degrees C. The numbers are not important, but the fact that temperature falls with height in the troposphere but not in the stratosphere is very important indeed.

Imagine that the Sun shines brightly and warms an area of ground surface. The air in contact with this warm surface is also warmed and, being warm, its molecules move more vigorously and the air expands. As it expands, it becomes less dense and because it is less dense, volume for volume it becomes lighter than the cooler air which surrounds it. Therefore it rises, and as it rises, it cools. A point may be reached at which its temperature and density are similar to those of the surrounding air and then it will rise no further. If it started out much warmer than the surrounding air, it may continue to rise all the way to the tropopause. By the time it reaches the tropopause, however, it will have cooled to the same temperature as the air above it. It must do so because the tropopause marks the boundary above which temperature remains constant with increasing height. Therefore the air will not enter the stratosphere.

Our parcel of rising air is entirely theoretical. In the real world such surface heating would produce not a single, distinct parcel of warm air, but a stream of rising warm air—a thermal current to prolong the delight of the pilots of gliders and hang gliders. The essential fact is that very little air from the troposphere enters the stratosphere. Where air is cooler than the surrounding air, and more dense, it will sink, but in exactly the same way, air from the stratosphere is inhibited in its descent by the tropopause, because just above that boundary, temperature remains constant with changing altitude. Air is exchanged across the tropopause, but the process occurs mostly above the equator, where air enters the stratosphere, and above the poles, where it leaves the stratosphere. The rate of exchange is slow and the two layers are quite distinct.

The first consequence of this is that of any material which enters the troposphere, very little crosses into the stratosphere, but that such material as is carried across the tropopause tends to remain in the region it enters. It is just as difficult for particulate matter to move from the troposphere into the stratosphere as it is for gas molecules, but once it has succeeded in doing so, it will remain in the stratosphere for a long time.

The second consequence is that the stratosphere tends to be very stable and the air within it very dry. It is dry because water vapor can enter the atmosphere only by evaporation from the surface of the Earth, and as the atmosphere ascends, so the water is condensed—we might almost say wrung—out of it. The temperature at the tropopause is so low that only a few parts per million of water vapor remain in the air and so there is very little water present at the boundary to cross into the upper atmosphere. The stratosphere is stable because it is protected against the violent events that make the troposphere a turbulent region.

The turbulence is caused by the exchange of heat between the surface of land and water and the air immediately above it. This causes the formation of large masses of air of different temperatures, which move against one another and mix only slowly. A warm air mass riding over a mass of cold air produces a weather front. Such air masses mix slowly because although air will flow from an area of high pressure to an area of low pressure, the rotation of the Earth deflects air that is moving to the north or south. Air, after all, moves with the Earth as the Earth rotates, but the actual speed of rotation changes with latitude—thus a point on the equator must travel a greater distance than a point at a high latitude in order to complete one revolution in a day, and so it must move faster. Air that moves north or south commences its movement while still moving at the rotational speed of the latitude from which it moves. In the northern hemisphere, if it moves north, the eastward movement it "inherited" will be faster than the eastward movement of the surface over the area it enters, and so it will move eastward. If it moves to the south, the deflection will be in the opposite direction. The result is to cause air to move around areas of high or low pressure in a shallow spiral, rather than directly to or from the center.

At the same time this is happening, the uneven surface of the land is exerting drag on moving air, is causing winds to eddy and swirl around obstructions, and is causing air to rise and cool—perhaps losing some of its moisture—as it crosses high ground and to sink and warm as it crosses low ground. Within the air masses themselves, differential surface heating and cooling cause local convection effects. It is this turbulence, this inherent instability, which produces our weather, and since water is involved and frequently changes its state back and forth among the solid, liquid, and gaseous phases, not only is the weather diverse but it is also very difficult to model accurately, so that the prediction of what will happen next is fraught with uncertainty. It is hardly surprising that some of the largest scientific computers in the world are devoted to the modeling of weather patterns.

The weather is confined entirely to the troposphere, however. The stratosphere influences strongly what happens in the troposphere, although its influence is not well understood, but it experiences none of the weather with which we are familiar.

Let us return now to all that dust that we may suppose entered the atmosphere following the planetesimal impact. As we said earlier, much of it will have been ejected violently into the upper atmosphere and some may even have been expelled from the atmosphere entirely. But some will have remained in the troposphere. Small particles that float in the atmosphere are known scientifically as "aerosols" and this is the word we shall use from now on to describe them.

The turbulence in the troposphere ensures that aerosols seldom remain there for long. Sooner or later they are washed out by the rain or carried by air movements into contact with surfaces, to which they adhere. Unless they are very large, they do not fall rapidly under their own weight: the viscosity of the air opposes the force that gravity exerts upon them. As they are lost from the troposphere, they are replenished from above. Despite the slow rate of exchange across the tropopause, there is some exchange and aerosols will continue to enter the troposphere until the stock of them residing at higher levels is exhausted.

Aerosols can affect the temperature of the stratosphere, but this does not necessarily produce a similar effect in the lower atmosphere. It all depends on the aerosols themselves. If they are very small—about 1 ten thousandth of a millimeter in diameter—most probably they will cool the lower atmosphere. If they are larger than this, they may have no effect or may bring about a warming.

The size of the aerosols determines the radiation wavelengths they will affect, and their color determines whether they will absorb radiation or reflect it. The atmosphere receives incoming radiation from the Sun, which is mainly short-wave, and long-wave radiation from the warmed surface of the Earth itself, because when a surface is heated it emits long-wave radiation. Very small aerosols can intercept short-wave radiation. If they reflect it, the radiation will be scattered, but because the source of the radiation is above them, much of the reflection will be back into space. Seen from space, the presence of many such aerosols will make the Earth appear to shine more brightly: it will increase the planet's albedo (reflective power).

If the aerosols absorb the radiation, it will heat them and so they will emit long-wave radiation, and because they themselves are the source of this emission, it will occur in all directions and some of it will be directed downward. In the atmosphere as a whole, the amount of incoming radiation is almost precisely equal to the amount of outgoing radiation. If it were not, the Earth would grow either warmer or cooler.

Stratospheric aerosols may either reflect radiation or absorb and reradiate it, but in either case, they accelerate the export of energy from this region of the atmosphere, and so the result must be a cooling, so far as the stratosphere itself is concerned. If they reradiate, however, a proportion of this exported energy will enter the troposphere. Some of it will be intercepted by tropospheric aerosols and that which is not intercepted will penetrate to the surface. It will warm the surface or the intercepting aerosols and cause further reradiation. Again, some of this reradiation will be lost into space, but some will be directed downward, to warm the surface again, and some will be intercepted by aerosols, which it

will warm. In this way, energy can become trapped in the troposphere.

In the troposphere, transparent or light-colored aerosols—those
with a high albedo value—will produce a cooling, and dark-colored
ones—those with low albedo value—will produce a warming. In
the modern world, the most common aerosols in the stratosphere
are of sulfuric acid, which has a very high albedo value. It is
formed in the stratosphere itself by the oxidation of gaseous sulfur
compounds which enter across the tropopause. In the troposphere
the most common aerosols are also sulfuric acid, but also am-
monium sulfate and soil—which is rock dust. These too have a high
albedo value and tend to cause a cooling of the surface. Soot, on
the other hand, has a low albedo value and causes a warming.

In an article in *American Scientist* (Vol. 68, May–June 1980, p.
268) Owen B. Toon and James B. Pollack summed up the present
state of our knowledge about the effect of atmospheric particles:
"Atmospheric aerosols can affect climate, certainly they have done
so in the past, and possibly they are doing so now. The kind of
effect depends on the size and composition of the aerosols and their
location—i.e., troposphere or stratosphere . . . . Adding nearly
transparent materials to the lower atmosphere, such as sulphates
and most soil particles, tends to cool the Earth's surface. Adding
opaque materials to the atmosphere, such as soot, tends to warm
the atmosphere. Since human activity is adding soot, sulphates, and
soil to the lower atmosphere in different regions, some areas are
probably being warmed and some are being cooled."

When we consider aerosols that enter the troposphere as a result
of a volcanic eruption or of the kind of impact we have described,
their most probable effect will be to warm the surface. Stratospheric aerosols, on the other hand, may produce either a cooling
or a warming of the ground surface.

This can be confusing because it seems to contradict common
sense. If you are out of doors on a day of hazy sunshine, you will
feel cooler than you would feel if the Sun were blazing down from
a clear sky. Your senses do not deceive you: you really are cooler.
The haze reduces the amount of direct solar radiation you are
receiving—by absorbing it itself, far above your head—and it is the
direct radiation that warms you. However, the haze forms a blanket that also absorbs heat that is being radiated back from the surface. You will feel cooler, but the atmosphere, indeed the environ-

ment as a whole, actually is warmer, because the radiant heat is being absorbed and distributed throughout a mass that includes the atmosphere all the way up to the top of the haze.

As though this were not complicated enough, molecules of carbon dioxide and water vapor also produce a warming effect but by a slightly different route. They are transparent to short-wave radiation but opaque to long-wave radiation. Thus they permit incoming radiation to pass, but outgoing long-wave radiation is trapped by them and then reradiated. This is called the "greenhouse effect," although in a greenhouse the "effect" is produced mainly by the glass, which is also transparent in the short wave and opaque in the long wave but whose more important effect is to prevent the circulation of air, trapping air that is being warmed but cannot be replaced by incoming cooler air.

A few years ago many people believed that by burning hydrocarbon fuels (coal, oil, and natural gas) that emitted carbon dioxide, human activity was tending to warm the Earth and by pumping aerosols into the air from factories and the worldwide expansion of agriculture, it was tending to cool the Earth, so that the two effects more or less balanced one another. We know now that this is not so. The climate of the northern hemisphere is cooling for reasons quite unconnected with human activity, and the man-made emissions of carbon dioxide and of aerosols are additive rather than balancing: both emissions tend to produce warming.

We wished to emphasize the complexity of events in the atmosphere, and especially of those involving aerosols, in order to show that the Alvarez idea of a global darkening is not necessarily contradicted by the fact that so far as we know no global cooling occurred as a consequence. We have no way of knowing the composition of the dust cloud that the planetesimal impact produced. Since a very great deal of the material from which it condensed entered the atmosphere as, or from, a plasma cloud, it is clear that we cannot regard this cloud as one produced from a kind of magnified Mount St. Helens. It was entirely different.

What is certain is that the release of so much energy into the atmosphere and the injection of so much material from the sea and from rocks would have caused a "pollution incident" the like of which we have never seen, and, it is to be hoped, never will see.

The atmosphere, you will recall, consists of about 78 percent nitrogen and about 21 percent oxygen. The atmosphere 65 million years ago had a similar composition. When air is heated strongly, the nitrogen becomes oxidized to form a series of nitrogen oxides. This happens, for example, when lightning delivers energy, it happens in hot fires, and it also happens in high-compression internal combustion engines. The governments of the world are concerned to reduce such emissions from gasoline-powered vehicles and from industrial furnaces.

We are supposing a huge release of energy and it seems certain that this would have produced very large quantities of nitrogen oxides. Bodies moving fast through air cause the oxidation of nitrogen. It has been calculated by Chul Park (in a NASA document issued in 1972) that the orbiting space shuttle might generate 10 tons of nitric oxide at each reentry, and he and Gene P. Menees calculated (*Atmospheric Environment*, Vol. 10, No. 7, 1976, p. 535) that a substantial proportion of the nitric oxide known to exist in the atmosphere is produced by meteoroids, most of it at around 95 kilometers (60 miles) altitude. Nitrogen oxides produce several effects. Nitric oxide is oxidized further and hydrated in the air to produce nitrates and nitric acid. Nitric oxide can also react to produce nitrogen dioxide in the presence of ozone. Nitrogen dioxide absorbs short-wave radiation, decomposing as it does so into nitric oxide and atomic oxygen. Present in large amounts nitrogen dioxide could exert a tropospheric cooling influence. Nitrogen dioxide is also a key component in the reactions that produce photochemical smog.

Nitric acid vapor can also form condensation nuclei for water vapor. It is washed from the air fairly quickly and on entering the soil in small amounts can provide nourishment for plants. But in large quantities nitric acid has less beneficial effects. These would have included an increase in the acidity of the soil and surface water. Today, emissions of sulfur dioxide from the combustion of high-sulfur fuels cause ecological problems and serious difficulties for farmers and foresters in the region of their fallout, and the fallout of nitric acid would have very similar consequences.

The ecological problems due to sulfur dioxide fallout have been popularized (and misunderstood) as being caused by "acid rain." The name is graphic, suggesting that atmospheric pollutants may be washed from the sky, increasing the acidity of rainwater and so the acidity of the soil. The confusion began because ordinary rainwater is weakly acidic, making it difficult to measure acidity contributed by pollutants. Clearly the soil was becoming more acid, but rainwater that was examined was not abnormally so. In the end it was discovered that the atmospheric pollutants are brought to the surface not by being washed out in the rain so much as by being deposited directly from the air on to the soil and plants. In the soil the acid may react with aluminum compounds, yielding further compounds of aluminum that are toxic to fish when washed into fresh water.

The effect of increasing the acidity of the soil is to injure many groups of plants that cannot tolerate acid conditions, and if the soil becomes sufficiently acid, it will be reduced to supporting only a heathlike flora. The extent of the differences in flora caused by the acidity of the soil or water has been studied. Acidity is measured on what is called the "pH scale" against water, which is neutral and has a pH of 7.0; pH values less than 7.0 indicate increased acidity, while those more than 7.0 indicate increased alkalinity. In a temperate latitude mire with a pH of less than 4.1, there may be an average of five plant species per square meter; where the pH is more than 6.0 there may be ten species. In open dry-land habitat, on slightly alkaline soils, there may be about twenty species per square meter; where the pH is less than 4.0 there are only two or three. In farmland, woodland, and wasteland the picture is the same: when the pH falls below 4.0 the number of plant species drops sharply. Among modern farm crops, barley and sugar beet are especially sensitive to increases in acidity.

In the Late Cretaceous the effect of increased acidity due to fall-out would have been to alter the flora, probably over very large areas and in regions that formerly had a neutral or alkaline soil, and flora to suit it; the result may well have been desert—a virtual absence of vegetation that would persist until either the soil recovered or until acid-loving species colonized it. Any substantial and enduring change in the floristic composition of a large area will have a serious effect on the animals dependent on that flora.

A large and sustained alteration in the acidity of the soil will have a less certain effect on the ecology of the soil fauna and microorganisms, acting mainly through changes in the purely chemical reactions that occur. Many microorganisms tolerate a wide range of pH, but some will die out. Those that can tolerate the new conditions will multiply to the limit of their food supply, and until migrant species adapted to the new conditions enter the soil, the number of species will be reduced. The soil will be impoverished ecologically and this, too, will affect larger plants whose relationships with soil organisms are close. A layer of matted, peaty, dank, dead grass lying above grassland is a diagnostic feature of an acid soil, and it is due to the failure of the bacteria that normally decompose dead grass. Although the grass itself tolerates a wide range of acidity, it cannot thrive unless old growth is cleared away to permit new leaves to emerge and photosynthesize.

Nitrogen-fixing species are very tolerant of low pH values, but they are very intolerant of nitrogen compounds introduced to the soil from outside sources. In the case we are considering there would be a considerable injection of nitrate and the acids themselves will contribute nitrogen. For this reason we may expect the nitrogen-fixing organisms to have been depleted.

So far the human contribution to the acidity of the atmosphere has been modest, but it has produced effects that are serious. In Scandinavia, where many soils are inherently rather acid, pollution from Britain, East and West Germany, and Poland has lowered pH values still further and has reduced the productivity of soils to such an extent that the governments of Sweden and Norway have been urging for years the adoption of some effective means of regulating such transfrontier pollution. In parts of Canada and the northeastern United States the natural flora has suffered severely from acid emanating originally from industry in the midwestern United States. This, too, has been the subject of governmental protests on both national and state levels. We are supposing pollution on a scale so much larger than any that has been experienced recently that we can but guess at its results. Were whole regions reduced to desert? It is possible.

Oxides of nitrogen would have been present in the stratosphere after the planetesimal impact, not only because they were transported across the tropopause, but because they were formed at high altitude just as they were at low altitude. Some years ago many people feared that the presence of nitrogen oxides in the stratosphere—in those days the potential source of them was the proposed fleets of supersonic commercial aircraft—would cause depletion of the ozone layer. The oxides, it was suggested, would engage in reactions involving atomic oxygen to form stable compounds and this "locking up" of atomic oxygen would inhibit the formation of ozone. In the end it was calculated that if such an effect did occur, it would be very small and unimportant provided the number of aircraft flying in the stratosphere remained small. More recently it has been found that in modest amounts nitrogen oxides do affect the chemistry of the stratosphere, but that they enhance the formation of ozone rather than depleting it. In any case, the size of the world fleet of supersonic transport aircraft has remained small.

What would have happened in the planetesimal event we are considering? We have no way of knowing. Our knowledge of stratospheric chemistry is improving, but it is still very far from perfect and in this case so many of the factors involved are unknown to us. It is possible that the ozone layer was depleted or that it was depleted in some places or at some altitudes but enhanced at others, according to the quantities of nitrogen oxides produced. In the end it matters little. In a moment we shall see that there would have been another, far more dramatic, chemical consequence of a large impact that almost certainly would have caused a substantial depletion of ozone, but that this depletion, if it occurred, was of no great significance. The complete disappearance of the ozone layer may be less serious than some environmentalists suppose. At the very least, if a planetesimal 10 kilometers in diameter collided with the Earth at full tilt, the loss of the ozone layer would be one of the more trivial consequences!

This picture is complicated further by the fact that the release of a large amount of energy in the atmosphere would have broken the bonds that hold atoms together within molecules. There would have been much atomic oxygen produced, some of it reformed as ozone, and so the event itself would have added ozone to the atmosphere. In the troposphere the ozone would have been yet another pollutant: to air-breathing animals, ozone is irritating in very low concentrations and in all but low concentrations it is intensely poisonous.

In saltwater—and the deposition of nitrogen compounds would have occurred over sea as much as over land—a sudden sharp reduction in the pH would have altered many chemical reactions. In our case a sudden and arbitrary change in the pH, together with the input of energy which we know raised the temperature of the water throughout the world, would have produced complex and probably deleterious effects, but most likely only local ones, for the buffering effect of the sea is very great. Substances must be deposited in it in vast quantities if they are to alter its chemistry more generally. Among the reactions to be modified locally would have been those involving carbonate and bicarbonate ions which will have affected marine organisms generally, but those with calcareous shells in particular. These organisms were killed in very large numbers, of course, at the end of the Cretaceous.

We should remember, too, that an increase in vulcanism throughout the world—and most especially in the North Atlantic region if the emergence of Iceland is to be considered as one result of the event—might have injected other acid-forming elements into the atmosphere, including in particular those of sulfur.

To understand the speculative chemistry just after the plane-tesimal impact we need to consider again the nature of the impact itself. Let us assume that the body entered vertically and that it reached the sea surface 10 seconds after its encounter with the outer edge of the atmosphere, at a height of 160 kilometers (100 miles). When it reached the surface, the body would have been pushing ahead of itself a few meters of highly compressed and in-candescent air. This layer would have contained nearly all of the air in the 10-kilometer-wide column of its passage. As it entered the sea, a similar zone of dense, incandescent plasma would have been thrust ahead of the body during the second or so it took to reach the sea floor. While this was happening, the front face of the plane-tesimal would have been heated and compressed similarly as the shock wave of its impact was transmitted upward. The final stage would have been the encounter with the bedrock of the sea floor. At this point the body would still have had at least half of its original energy and be traveling at close to Mach 40.

Rock would now have been added to the sandwich of com-pressed plasma made from air and sea. Eventually, at some consid-erable depth—at least 16 kilometers (10 miles)—the impact force would have been matched by the immense pressure of the plasma, heated now to a temperature well above that of the solar surface. Nothing solid would now have remained of the planetesimal itself, but the dense plasma cloud that replaced it would have expanded explosively upward, thrust by the vast pressures exerted by the air-sea-rock plasma sandwich beneath it. The object would have ap-peared to bounce.

The elapsed time from impact to rebound would have been no more than a few seconds. During this time the seawater and sea floor would barely have started their long, and by comparison slow, journey out to the far reaches of the crater that was destined to be formed and whose edges would have been nearly 200 kilometers (124 miles) apart. The sea and rock moved to form the crater would have occupied 60 times the volume of the planetesimal itself. If their motion was at the speed of sound, then it would mean that some 30 percent of the impact energy would have been expended in forming the crater, the rest of it converted to the heat and motion of the plasma except for some lost overcoming drag during the passage through sea and rock. The energy lost during the planetesimal's passage through the air and in magnetohydrodynamic effects seems unlikely to have amounted to more than a few percent.

Normally bodies that enter the Earth's atmosphere from space to impact at the surface do not cause the ejection of material back into space. Were they to have done so in the past on Earth, they are even more likely to have done so on the Moon, where the escape velocity is lower and much less energy is required to eject material into space. We should expect, therefore, that were past impacts to have caused the ejection into space of lunar material, then some of that material would have reached Earth. Yet no meteorites of lunar origin have ever been found. However, the Earth is unusual in its covering of ocean and it could be that the rebound of a planetesimal from 5 miles of ocean compressed to a plasma would project at least some of the remnants of the body into Earth orbit.

Let us leave for a moment our consideration of the pollution of the atmosphere and speculate about the fate of marine organisms; and let us think not about the open ocean, which is deep and sparsely populated by living organisms, but about the much shallower waters over the continental shelves, which are densely populated. Whatever else may have happened, very few living things could have survived in the North Atlantic and any survivors would have been living in the mud on the seabed. Those inhabitants of the upper waters that were not vaporized, whose contribution to subsequent proceedings must be considered in relation to the atmosphere, would have existed as corpses. What happened to those corpses?

When plants and animals die, their remains are consumed by other plants and animals in a hierarchy that ends (as it began) among the microorganisms. Thus, complex organic structures are decomposed to simple inorganic compounds that reenter the nutrient cycle. In this case, though, much of the cycle would have been destroyed. Survivors among the microorganisms would have multiplied rapidly amid an abundant food supply, so simple nutrients would have been available in plenty, but it seems doubtful that there were plants to use them or animals to graze the plants.

In deep water the corpses would not have drifted gently to the sea floor, to be buried reverently by the slow process of sedimentation, but in shallow water they would have done, and apart from those carried into the sea that was Denmark, they would have decomposed. Like the atmosphere, deep water is stratified. There is a temperature gradient, as there is in the air, and a level at which the temperature remains constant called the "thermocline." The thermocline forms a boundary that is no easier to cross than the tropopause. Material does cross it, and deep-sea species depend on the manna from above with which it keeps them supplied, but the crossing is slow and before dead organisms pass they may be carried very long distances by ocean currents. We can estimate the transport of such organisms in the modern ocean, because the movement of currents is well known, but the prehistoric North Atlantic was a different size and shape, and following the planetesimal event its waters would not have behaved as they do in normal times. You will remember that a large quantity of water was removed on impact from the impact site. This water would have been replaced, and it could only have been replaced by water entering from the south. In the North Atlantic of 65 million years ago, that is the only direction from which it can have come. It is likely that the inrush of water would have created some kind of circulation pattern, so that the contents of the North Atlantic—simple nutrient compounds and corpses—would have been mixed thoroughly. This might have made it easy for decomposition to turn much of the water, especially in coastal and estuarine areas, into a rich nutrient "soup."

The nutrient soup would not have remained unused for long. Simple plants—algae most likely—would have arrived to exploit it. In some places at least, there may have been large-scale algal blooms. When the algae died they, too, would have been decomposed.

The process of decomposition consumes dissolved oxygen from the water—the organisms involved require oxygen for respiration and such decomposition is essentially the oxidation of the carbon—and those organisms that are critically dependent on oxygen-rich water die from asphyxiation. This process of nutrient enrichment is known as "eutrophication." Once it begins, it can continue until the extreme situation is reached in which the water supports very few species other than bacteria. As species die out one after another, their remains contribute to the eutrophication. In this case, the population of the upper waters would have been killed first, but eutrophic water could have inhibited their return.

Can this have happened? Can the sudden death of countless billions of microorganisms, plants, fish—and reptiles, of course—have led to the widespread eutrophication of coastal waters and estuaries around the North Atlantic? It is possible.

There is a secondary effect associated with the eutrophication of seawater. Certain of the planktonic organisms produce powerful toxins, and from time to time such organisms are known to bloom in eutrophic waters. Thus the eutrophication of seas might have been accompanied and accelerated by the poisoning of organisms. This occurs directly, with the injury or death of susceptible species that consume the toxic species, and indirectly, by the concentration of poisons in the tissues of species that are not susceptible themselves but which are then eaten by species that are susceptible. Many mollusks, for example, can tolerate the poisons, but some species that eat the mollusks cannot, and die. In effect, this would have rendered a large amount of water toxic to most larger animals and possibly to some planktonic species. Might this water have mixed with water elsewhere to spread the effect?

There is a final effect that involves both sea and air, and as we promised, it is the most dramatic of all.

We must return to the energy of the impact. The energy released when a large—say 200,000-ton—ore-carrying ship traveling at 1 knot collides with a dock is 20 million joules. (The joule is the international standard unit of work. In electrical terms it is the energy dissipated by one watt in one second.) You can see and hear the impact, but if the cargo is not inflammable, usually there is no fire, smoke, or great emission of heat. The explosion of 1.2 kilograms (2.64 pounds) of TNT yields the same amount of energy, but because it is released in a much smaller volume, the effects are more dramatic.

To say that the energy yield of our planetesimal impact would have been equivalent to the explosion of 100 million megatons of H bombs is true but misleading. The impact would have been 36,000 times faster than that of the ship collision, but the rate of energy release much less concentrated than in an H bomb explosion.

Energy intensity can be expressed as joules per kilogram. For the ship collision it is 0.1; for a chemical explosive such as TNT, 17 million; for the planetesimal impact, 180 million; and for the nuclear explosive it is 200,000 billion. It is as though each particle of the planetesimal had been made from an explosive 10 times more powerful than TNT, but about 1 million times less powerful than a nuclear explosive.

These distinctions are important, for the type of damage from an impact depends greatly on the rate of release of the energy and its intensity. The ship, like a bulldozer, is a powerful destroyer of solid structures, but harmless to the molecules of the air. The nuclear explosive is powerful enough to fuse the very atoms themselves. Our meteorite, 10 times more powerful than TNT, would have had energy enough to reduce molecules to atoms and atoms to electrically charged ions, but not enough to alter the atoms themselves. The capacity of the impact to sever molecular bonds is important to our understanding of the nature of the damage done, for such severence and rearrangement of molecular bonds, when applied to normal constituents of the air and sea, can give rise to new and very toxic combinations.

We know that the highly compressed interface between the planetesimal and the sea would have risen to temperatures much hotter than the surface of the Sun. We know that at such temperatures every chemical bond is severed and everything is stripped down to bare atoms, electrically charged ions, and free electrons. What we do not know is how much of the sea would have been transformed into such a plasma, nor the rate and manner of its mixing with the air.

If we guess that 10 percent of the impact energy would have been used to convert seawater to incandescent gas at 3,000 degrees C, this would correspond to about from 100 billion to 1,000 billion tons of sea. On first mixing with an equal mass of air, the first set of products undoubtedly would have included chlorine atoms, for 2 percent of seawater comprises chlorine ions. Whether or not these remained free as chlorine or recombined with the sodium and hydrogen of the incandescent cloud would depend on the manner of the cooling of the plasma and on how rapidly it mixed with cooler air or seawater. If only 1 percent of this chlorine had remained free as chlorine gas, it could have amounted to as much as 200 million tons of chlorine, enough to turn a great deal of ocean into a sterile, algae-free swimming pool. The affected sea might have been an area of 40,000 square kilometers (15,500 square miles) to a depth of 5 kilometers (3 miles) or an area of 1.96 million square kilometers (757,000 square miles) to a depth of 100 meters (328 feet), for example.

Fortunately, the ubiquitous presence of methane in the air and sea would have ensured that the chlorine could not travel far through the air. Methane reacts rapidly with chlorine in sunlight and there is enough methane in the air to react with 30,000 million tons of chlorine. It might have been that in the dimness caused by the dust, the filtered light of the postimpact period would not have promoted this reaction, but nevertheless it is very unlikely that chlorine could have traveled from the hemisphere of impact to the other side of the world.

We have introduced this speculation about chlorine to illustrate how the vast quantities of crust and sea disturbed by the impact and heated to incandescence could have produced pollution global in scale. Free chlorine would by no means have been the only plausible product. Other compounds of the element chlorine, such as hydrochloric acid and hypochlorous acid, are also possible. We do not know if such poisoning occurred but it will be interesting to follow the evidence from the field investigators to see if there are signs that it did.

In his interpretation of this catastrophe, Kenneth Hsü wondered if poisoning, global in scale, could have come from the impact of a comet bearing ice rich in hydrogen cyanide (HCN). It is easy to calculate that if 10 percent of a comet head 10 kilometers in diameter were HCN and if all of it survived the impact, then the top 100 meters of sea would contain 1 part of HCN to 100,000 parts of water. Probably this would have been enough to disturb or eliminate many, but not all, of the species present.

It is very unlikely that more than a few percent of the HCN would have survived the fierce heat of the impact and the oxidation that followed. As with chlorine, HCN is fairly reactive and

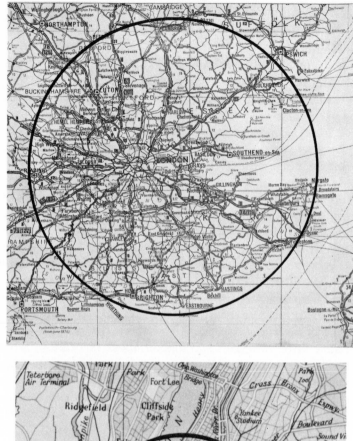

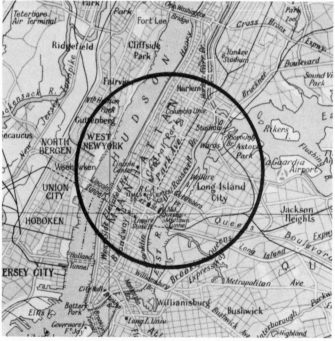

Top A map of south-east England shows the size of crater which might be blasted out by the impact of a planetisimal of the size thought to have caused the Great Extinction. *Below.* A map of New York shows the size of the planetisimal — about the same as Manhattan island.

would have been very unlikely to have penetrated in harmful quantities across the hemispheric boundary.

What can become toxic and is known to have been distributed worldwide after the impact is osmium. Normally, this noble element is inert and harmless, but when it is heated in air, it oxidizes to form the intensely poisonous vapor osmium tetraoxide. It would be an irony if one of the very elements that led the Alvarez team to the planetesimal cause of the extinctions was itself the agent of those extinctions.

Extraterrestrial matter is distinctly unhealthy in many ways, quite apart from its burden of osmium. Compared with the Earth's crust, it is enriched with arsenic, lead, copper, antimony, selenium, chromium, and barium. All of these are toxic elements and they would have been dispersed in vast quantities all over the world by the planetesimal impact. Among all of the other insults suffered by life as a result of the impact, it would have been dosed with a cocktail of poisons that would have won a seal of approval from the Borgias themselves.

Compared to this, our own efforts at industrial pollution—indeed, of any efforts we can imagine ourselves making—are puny indeed. The miracle is not that species became extinct, but that any survived. Yet they did survive, and once the event was over and done with, life reasserted itself with new vigor.

9

The Scene of the Crime

Not very long ago a motor manufacturing company, which shall remain nameless, proclaimed the merits of its latest product in the British national press by comparing rival cars with dinosaurs. The dinosaurs, the advertising copy said, were cumbersome, heavy, unsophisticated, slow, and could not hope to compete against more advanced rivals. Happily, there were immediate and sharp protests from people whose knowledge of those antique animals was considerable and the advertisements were withdrawn. The description in the advertisements was so inaccurate as to be offensive to those who have devoted years of their lives to studying extinct species and to speculating, on the basis of the best information available to us, about how they may have lived.

What is more to the point, the description did not, cannot, make sense. Living organisms do not thrive on our planet, or presumably anywhere else, unless they are adequately prepared for the conditions under which they live. The fact is that the large reptiles that walked the Earth before and during the Cretaceous were highly successful animals. Theirs was a world to which they were adapted supremely well.

The other fact is that the advertisement did no more than perpetuate a popular myth. It is time for us to dispel that myth, for unless we do so, there is a risk that no matter how convincing the evidence for a catastrophe may be to us, a suspicion will continue to lurk in the back of some minds that species become extinct because they are "poorly designed."

Of course, it is scientific heresy merely to use the word "design" when speaking of biological organisms and yet we are more prone than we may wish to believe to entrapment by cultural concepts with which we were indoctrinated as small children. It was as children, after all, that most of us first heard of those marvellous monsters of far-off times. Is it possible that at least some of us continue to associate them with the rather questionable moral precepts with which our tales of monsters usually are surrounded? At least one team from an advertising agency appears to do so, although we may hope that the team has now been rescued.

Perhaps the trouble began about a century and a quarter ago, when information about large extinct reptiles combined with information and ideas concerning the evolution of species became hopelessly mixed up with nineteenth-century theories of economics and attempts to understand the way industrial societies work. There was deliberate intellectual dishonesty, as some theorists sought to apply to human affairs scientific concepts borrowed from quite different fields and so to justify the political systems they preferred. It is hardly surprising that the versions of science which reached ordinary people, and more especially ordinary children, were muddled. Some versions had been filtered through the popular press. Others had been processed by popularizers who drew moral conclusions where no such conclusions could legitimately be drawn. Through it all, ideas about evolution and so about extinct species became associated in most minds—for the scientists themselves were not always immune—with the concept of "progress."

It may sound remote, distant in time from our enlightened age, but if, in some second-hand bookshop, you should come across books about prehistory written for children only half a century ago and on bookshelves in many homes much more recently than that, you are likely to find essentially nineteenth-century attitudes leaping a good third of the way into the twentieth. Our new attitudes, our new knowledge, really are rather new as far as many people are concerned. Should you doubt this, it is worth remembering that although the concept of evolution, in the sense of the mutability of species, was accepted widely by the end of the nineteenth century, the Darwinian idea of natural selection as the principal driving force for speciation presented such profound difficulties that it was not accepted fully by the scientific community until the 1940s.

We should deal first with anthropomorphism, the word which sends cold shivers down the backs of zoologists, and a view of the world that is more prevalent, and more difficult to avoid, than even they may suspect. We are human beings. We see the world as human beings. Information about it reaches us through human senses and is interpreted by our human brains. We can see that other animals possess similar senses. They have eyes and ears, tongues that taste, noses that detect odors, and skin that is sensitive to touch. Some of the things that cause us pain seem to cause them pain, too. What could be more natural than to suppose that things that we find pleasant, they, too, find pleasant and, leading directly from this, that they experience emotions similar to our own? Indeed, animals do display what we describe, for want of better words, as "anger" and "fear." Why should we deny them more enjoyable emotions?

By seeking to identify with nonhumans in this way in fact, we are regarding them as humans, albeit of different outward appearance. We are interpreting their behavior in the light of our own experience and so projecting some part of our own humanity on to them. Being human, it is almost impossible to avoid anthropomorphism consistently, especially in our attitude to our domestic pets. Everyone has met the cat which "understands everything you say," and many children's books actually teach anthropomorphic attitudes in order to inculcate a respect for animals in young people. Sadly, overzealous attempts to counter anthropomorphic attitudes sometimes lead to an extreme view of animals as being nothing more than machines, incapable of enjoying pleasurable sensations or of suffering. In turn, this merely fuels the oversentimental view of nonhumans until the two extreme attitudes collide, often with marches by placard-carrying demonstrators whose vaunted pacificism masks a propensity for violence at least as great as the violence they oppose.

The inevitable consequence of anthropomorphism, and the aspect of it which affects our attitude to the Cretaceous reptiles, is that we cannot avoid comparing other species with ourselves. This is necessarily to the disadvantage of the nonhumans, for their lack of human attributes is held against them as a mark of their deficiency, often their moral deficiency. The polecat is "dishonest," the fox "crafty," the cat "cruel." When we add to this misguided view of the world the concept of evolution as progress that leads to ourselves, we add a new dimension. We are intelligent—according to the definition of "intelligence" we have devised to describe ourselves—and by comparison, reptiles are stupid. Any characteristic we possess is a characteristic we associate with success and its lack imposes a disadvantage. Therefore, since the dinosaurs became extinct, clearly their extinction must have been due to their lack of features which humans possess. They failed because they were unlike us; we are the successes by which their failure is measured —and we would ask you to note the ease with which use of the word "success" permits us to introduce its antonym. Emphatically, the dinosaurs did not "fail." Species do not fail. The concept is inappropriate.

The fact that it is only animals which we regard in this way— despite some attempts to talk to plants and otherwise to treat them as though they, too, were human—reveals a second and deeper error. Whether our grandparents admitted it or not, they were and we are animals, and we take it for granted that animals are more important in the great scheme of things than plants and that plants are more important than the myriads of microscopically small organisms of which we are hardly even aware. The overall attitude can be summarized quite briefly: "bugs" are dirty and make you ill and so are to be destroyed where possible; most plants are attractive and many are useful and so are to be cultivated with care; animals are like us, although some of them may be dangerous enemies. We have said before, and we say again, that this entirely natural view of the world is distorted. Indeed, it could not be more wrong. It is the "bugs" that operate the processes by which the planet is maintained as an acceptable environment for living organisms. If the disinfectant that "kills 99 percent of all household germs" lived up to its claim and if it were applied rigorously, we would be in a very sorry state.

Let us return now to that old nineteenth-century muddle. Charles Darwin was understood to have said that competition among individuals and species ensured that only the fittest survived. The fact that this is not quite what he really said is irrelevant. It is what he was heard to say. "Fitness," to many nineteenth-century European and American males, was a word that went along with "virility," "manliness," "athleticism," and "virtue" and with such concepts as physical strength and military prowess. "Survival" sounded very like "victory in battle" or "success in business." So "social Darwinism" was born and did a great deal of harm. When the popular idea of Darwinism was applied to nature as a whole, those who could accept that man is related evolutionarily to other species never doubted that we are the culmination of the process, that man is fitter than any species has been in the past. After all, we are here and they are not. The case hardly merits a hearing. Those who did not accept the relationship had no problem. Man is the lord of the Earth.

The use of the word "evolution" contributed to the confusion. As Stephen Jay Gould has pointed out (in *Ever Since Darwin*, New York: Norton, 1979) Darwin himself used the word seldom and with much caution, for in nineteenth-century England it was already in common use and it was virtually synonymous with "progress." The latest edition of the Concise Oxford Dictionary gives this as first definition of "evolution": "Opening out (of roll, bud, etc. . . .); appearance (of events etc.) in due succession; evolving, giving off, (of gas, heat, etc.)." It is not until the third definition that the biological sense of the word is given. Evolution, then, implies progress, a movement from somewhere to somewhere, with the probability of a goal even though the goal remains hidden. Darwin used the word reluctantly because although it might help to describe one aspect of the process, in a more general sense it contradicted flatly his view of the way that process occurs. Today the word is used widely, but it remains as scientifically inaccurate as ever and we would do well to bear this in mind. Any idea of progress, except in the limited sense of events, speciations if you like, occurring one after another in a series, is entirely wong.

To our Victorian forefathers, "progress" implied a concept with which the man of affairs could identify. When told that species had evolved from simple beginnings to organisms of great complexity, he could add this information to his own observation that on the face of the Earth human beings have the power of life and death over other large animals and over plants, and it became self-evident that we ourselves are the culmination of the process. We are the "dominant" life forms, lords of all we survey, and we can supply a little apparent respectability to the notion by claiming to be the most highly evolved of all organisms. As a matter of fact, "highly evolved" is a term that has no meaning whatever, but it sounds very grand.

If today we are dominant, then it would be churlish to deny that in former times, when we were absent, other forms were dominant. At one time it may have been jellyfish, at another time fishes, for a very long time indeed it was the simple single-celled organisms from which we are descended, and for a particular time it was the reptiles. On the battlefield of life, in the factory that is the world,

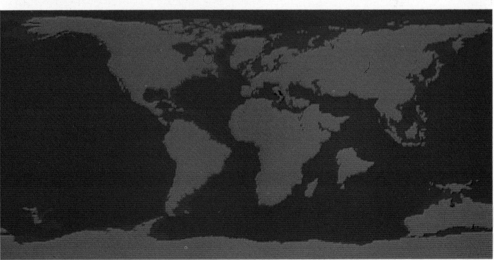

Top. The reconstructed fossil skeleton of a plesiosaur unearthed at Peterborough in 1970. Members of the order of Sauropterygia, the plesiosaurs were marine reptiles, not true dinosaurs, but they also died in the Great Extinction. (*Imitor*) **Below.** Map of the Earth 65 million years ago. The Atlantic is much narrower. The Middle East is not yet connected to the main body of Asia, and India is a separate giant island. (*Dr. S. Gull and J. Fielden/Science Photo Library*)

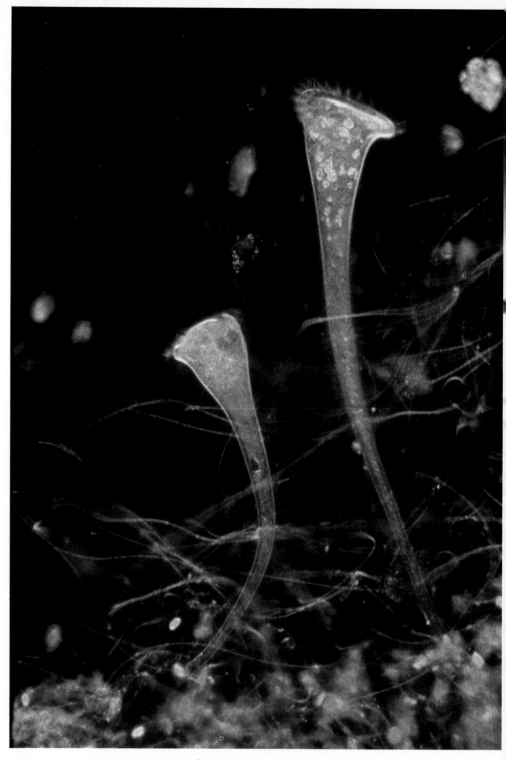

While larger life-forms fall victim to catastrophic events, micro-organisms such as these pond-dwelling, trumpet-mouthed Stentors survive and diversify. This pair, magnified 300 times, is seen amongst filamentous bacteria. (*John Walsh/Science Photo Library*)

only the dominant counts. By definition, he who is not dominant must be subservient, of less consequence. From time to time, however, there is a revolution. The factory owner is dismissed from his boardroom, kicked out of his mansion, and a new, more successful owner assumes power to become the new dominant. That is as it should be, for eventually it allowed our species to emerge. We are the proof of the moral soundness of the system.

So deeply has this Victorian (totally muddled and incorrect) picture of evolution influenced us that it is difficult to rid ourselves of it. Most people still think that mammals are the "dominant form of life" on the planet at the present time and that, among mammals, human beings are preeminent. Even distinguished scientists, who know better, allow themselves to fall into the common idiom when speaking of the Cretaceous Period and talk of "the rule of the dinosaurs." They and their colleagues know what they mean, but such loose talk perpetuates the popular confusion.

Dinosaurs did not "rule" and neither do we. Neither they nor we nor, for the matter of that, any other large plant or animal has ever "ruled." Kings and queens rule over human subjects. Animals and plants do not rule over other plants and animals. Nor is it true that we, the dinosaurs, or any other large plants or animals have ever been "dominant" as life forms. All we can say of them is that compared with other species of about their size they have been numerous and successful at reproduction. At the present time human beings are reproducing successfully—some would say too successfully—and we have colonized much—but by no means all and in terms of area probably not most—of the surface of the planet. We are not the most numerous of mammals, though. The mice outnumber us and so do the bats, not to mention the rabbits. It is true, of course, that we are by far the most successful of the manlike mammals! If we wish to use the word "rule" in its proper sense and list the biological organisms whose influence on the planet is such that their removal would produce chaos—anarchy— then that list must be headed by microorganisms and, among them, by the prokaryotes, organisms so small that only the more powerful microscopes can make them visible to our human eyes. Unfortunately, so far as our vanity is concerned, the list should go on to include the green plants, which use solar energy to manufacture sugars, and the fungi, which play an important role in the early stages of the decomposition of organic remains, but the animals—

the consumers of food made by microorganisms and plants—do not figure at all in the list. They are passengers so far as the system is concerned and it could function perfectly well without any of them.

What happened in history was that microorganisms, which may (or may not) have appeared first in shallow seas, so modified the terrestrial environment as to make it more favorable for themselves. Some of them, by means about which we can only speculate, began to collaborate and so multicelled organisms appeared. The collaboration became fixed, so that cells specialized within the communities that were the larger organisms, and those communities or organisms began to prosper. As they prospered, they modified their immediate environment further. Among them, random and extremely frequent small changes in the highly complex nucleic acid molecules that provide templates for the assembly of proteins, and which are inherited by offspring, caused individuals to differ one from another. Those individuals that were best able to feed and reproduce left more descendants than those that were less well endowed. With our Victorian vocabulary we like to think of this as "the survival of the fittest." It might be more accurate to call it "the survival of the average." Biological systems tend toward stability.

The reason they do is fairly obvious once we free ourselves from all ideas of progress and consider species not in isolation but as members of communities consisting of many species. A field on any farm will serve as a familiar example. Using the warmth of the Sun to provide the temperatures at which necessary chemical reactions can occur, microorganisms subsist on minerals which they obtain from small fragments of rock that has been shattered by the repeated freezing and melting of water. Other microorganisms subsist on the wastes the first microorganisms produce and some prey upon them. Certain of these organisms collaborate by living together within a shared cell wall and some include within the wall organisms that are able to use sunlight to manufacture sugars from carbon dioxide and water. They modify the atmosphere and so regulate, within broad limits, the temperature of their environment. Larger organisms, which some biologists believe arose from the further collaboration of small organisms, are able to feed and reproduce, but by doing so they modify the environment further. Principally, they provide wastes on which one group of organisms may feed and they themselves provide food for other groups. Without the microorganisms there would be no nutrients for large plants. The grass would not grow. Without the plants there would be no herbivorous animals. Without the herbivores there would be no carnivores. Each group exploits other groups and is exploited. When the farmer comes along and plows the field to sow his crop, he is behaving no differently from other organisms.

The important, but subtle, point to remember is that the success of one group does not and cannot imply the failure of the groups it exploits. Within an inherently stable system, an increase in one place can be sustained only by proportional increases elsewhere. Nor does evolution consist in the alteration of one kind of organism into another, somehow better, kind of organism, but of relatively minor changes and adjustments within the overall pattern, whereby living material organizes itself, now this way, now that, while the pattern itself continues. The complexity of individual organisms—arrangements of living material—may increase, but the simpler forms do not necessarily disappear. The microorganisms from which later species are believed to have arisen have not been supplanted and cannot be. They are still there and they are fundamentally important. Because species are so utterly dependent on one another, together creating the conditions under which all of them live, the system they comprise has very strong self-correcting mechanisms—negative feedback. The system is stable and tendencies to change are corrected provided they do not exceed the capacity of the system and so overload it.

New species can emerge and establish themselves only where resources exist which they can exploit, and in most systems for most of the time such resources do not exist. Everything will be in use, all niches will be occupied, and innovations will be doomed to fail.

We should expect, therefore, that major evolutionary change is most likely to result from a disturbance of stable systems and that the disturbance will come from outside the systems themselves. Probably there are exceptions, for even stable systems can break down, but when we seek the cause of major change the first place to look is outside the system rather than inside it.

Until very recently many attempts to explain change in biological systems based themselves on ideas of change within the systems. In the 1960s an eminent professor published a book on prehistory in which he used such an explanation, one which was very popular in the last century, to account for the disappearance of the dinosaurs. He called the process "phylogeronty."

The theory is that just as individuals have life cycles, so do species and groups of species. They emerge, multiply, prosper, but then they go into decline and finally disappear as the result of something very similar to senility. Because natural selection eliminates individuals and strains which are in some way weaker than others, during long periods of stability the effect is to reduce the size of the gene pool and so to increase inbreeding which weakens the population as a whole. Among the characteristics each generation inherits from its parents there are many that are adaptive, that help the individual in its daily life. But there are some that may become hindrances. Over many generations, says the theory, these "bad" characteristics accumulate until eventually they lead to reduced reproductive potential. The individuals possessing them, and this means all individuals of the species concerned, produce fewer offspring in each generation until they die out completely. They have become senile, decadent, and "evidence" cited in the case of the dinosaurs includes their alleged great size, slow movement, and an inability to adapt to changing conditions.

A rather similar idea, also popular in grandfather's time, complements this. It holds that some characters are able to develop in an uncontrollable kind of way so that they become more and more exaggerated in each succeeding generation. Thus the dinosaurs grew bigger and bigger, slower and slower, and more and more stupid until—as we were told when we were children—the brontosaurus (now called the "apatosaurus"), for example, was so slow and stupid it took the brute half an hour to respond to the dropping of a boulder on the tip of its tail. People who accepted this theory—and very many people did until quite recently—believed that the Irish elk, the largest member of the deer family that has ever lived, became extinct because its antlers grew larger and larger in succeeding generations until the day came when the stags became entangled in the vegetation, could no longer feed or even lift their heads perhaps. In the same way, the saber-toothed tiger, it was alleged, grew teeth that eventually became so large the poor animal could not close its mouth and so could not bite anything. Such examples show how nonsensical the theory is—and the phylogeronty theory is no less nonsensical—yet it persists in a slightly modified form. It is quite impossible for evolutionary processes to lead to animals that, as individuals, are maladapted. They could not live long enough to reproduce and so would be eliminated at once.

Many people still believe that species may evolve into more and more specialized forms which carry them into dead ends where they are trapped should the limited way of life for which they are adapted become unsustainable. The tiger is sometimes described as such a specialized animal on the grounds that as a large carnivore it needs a large range and so cannot survive if the area available to it is reduced. All this means, of course, is that the tiger cannot survive without food. Neither could we, yet we consider ourselves to be among the least specialized of species and consequently among the most evolutionarily robust. The list of highly specialized species of microorganisms, plants, and animals is very long, and many of those species have inhabited the Earth for hundreds of millions of years. Their specialization seems to have done them no harm, they seem not to have grown teeth, antlers, or other appendages so large as to incapacitate them, and despite their long ancestry they show no sign of senility. We can see, though, how the motor manufacturer came to write his advertising copy.

We claim, then, that the large Cretaceous reptiles were perfectly "fit" in an evolutionary sense, that there is no reason to suppose them to have been in decline. Now, what do we mean by "fit"?

Having rid ourselves of any idea of evolution as progress, we must ask ourselves what it is that marks evolutionary success. Obviously, a species that is numerous is successful, but just what is it that "succeeds"? After all, apart from those microorganisms that reproduce by cell division (mitosis) and so may be considered immortal, all individuals must die. Death is hardly an attribute of success.

The first clue to understanding the process was supplied, wittily, by Samuel Butler (1835–1902), who said "a hen is only an egg's way of making another egg." The egg carries from generation to generation the coded instructions that direct the construction first of the chicken and then of the new egg. Is it, then, the survival of those instructions which is all that matters? The idea that it is DNA—the replicating nucleic acid by which templates for the compilation of amino acids are transmitted from generation to generation—which is the subject of biological history and that evolution may be seen as the development of a variety of "survival machines" to ensure the immortality of particular configurations of DNA, has been argued most persuasively by Richard Dawkins, who summarizes his view in a popular and most readable book,

The Selfish Gene (London: Oxford University Press, 1976). The plausibility of his idea is enhanced considerably by the fact that cells contain much more DNA than is necessary for the construction of more cells like themselves, its only function apparently being to replicate itself and be passed down the generations. When we look at organisms, therefore, we should consider their "fitness" in the light of their success in ensuring the survival of the DNA they carry with them. They may carry this DNA into new habitats, thus spreading it more widely about the Earth, or they may transmit it to their descendants with greater efficiency than their predecessors.

There are two strategies open to a species. In one, they may produce many offspring. Most will die, but should the environmental conditions under which they live improve suddenly there will be large numbers of individuals available to take advantage of the improvement. Such a strategy is typical of species whose environment is subject to rapid and unpredictable change. The alternative is to produce few offspring so that the number of individuals is suited to the carrying capacity of the environment, which typically is rather stable. This is the strategy we see among primates, for example, as well as many other animals. In general, it is a strategy found more commonly in low latitudes and stable conditions, while its alternative is found in strongly seasonal environments in high latitudes, but there are many exceptions to the rule.

In examining the fitness of the reptiles, let us look first at their reproductive efficiency. The divergence of lines of individuals into what we recognize as distinct species requires that genes be mixed by the mating of individuals. While each individual possesses a complement of genes that are different in detail from those of any other individual, among particular individuals within a group the genes are sufficiently similar to permit them to pair with one another and so develop into a new individual also capable of reproduction. Because mating involves the pairing of genetic material, one strand from each parent, to form the complete double strand, it is necessary that the two strands "key" together chemically. For this to be possible, the strands may differ from one another only within certain limits. Thus one human, for example, may mate successfully with any other human (of the opposite sex), but a mating between a human and a nonhuman would fail because the two strands of genetic material would be unable to unite owing to substantial differences in the arrangement of the units comprising them. Such discrete groups of organisms within which reproduction is possible are what we call "species," and reproduction by the mating of individuals appeared first among microorganisms.

Later species laid eggs and the line which led (in time, not from the point of view of progress toward a goal) to the amphibians, reptiles, birds, and mammals included the fishes. Female fishes lay eggs that are fertilized by the males outside the body of the female. The eggs hatch to produce larvae, which must begin almost at once to fend for themselves. The fertilized eggs are liable to be eaten by other species—although in some groups of fish the eggs are defended—and most of the larvae perish in the jaws of predators. To ensure the survival of a requisite amount of DNA, each female fish must lay vast numbers of eggs—often hundreds of thousands. The procedure is wasteful of resources and most especially of DNA, but it works well enough for the fish. Their environment is subject to sudden change, in the amount of food that is available and in the degree of predation to which they are subjected.

Amphibians reproduce in much the same way, although in some of them eggs and even larvae may be retained in or on the body of one or the other parent, so that the chance of survival is increased for each individual offspring. However, amphibian eggs, like fish eggs, must be laid in water and the larvae that hatch from them live by swimming freely in water.

In reptiles the chance of survival for the young is increased further. Reptiles lay eggs with an "amnion." The amnion is a sac filled with fluid inside which the embryo can develop. Instead of swimming freely, prey to the many carnivores that inhabit rivers, lakes, and the sea, the entire aquatic larval stage in the development of the animal can be undergone within the egg and the egg hatches to reveal an animal very similar to its parents.

In some reptilian species this development is taken further. There are reptiles (the common British lizard is an example) the female of which retains the eggs inside her body until they hatch. They are separate from her body and obtain no nutrition from it, but ovoviviparity—the scientific term for this characteristic—provides protection for the egg and, apparently more important, an optimum environment for incubation. (There is some confusion of terms. Ovoviviparous species produce young that are alive when they leave their parent's body and consequently ovoviviparous species are often referred to as "viviparous," which strictly speaking is incorrect.) Such animals produce far fewer young, but usually they inhabit environments that are stable or whose vagaries can be predicted.

In mammals the egg becomes an integral part of the female body. Nutrient is supplied to it not from a store—the yolk—but directly from the body of the mother, and waste products are removed through the mother's own excretory system. Not only is the developing embryo fully protected from physical danger, but it is also protected nutritionally. During gestation its chance of survival is greater than that of a reptilian, amphibian, or fish embryo, but although it does not have to go through a larval phase the mammalian infant is born more or less helpless. It is incapable of surviving independently of its mother and in some species—including our own—two parents usually need to collaborate in caring and providing for it. Nevertheless, the mammalian arrangement does permit a rather higher proportion of young—of invested DNA, if you prefer—to survive than is the case among reptiles.

The colonization of dry land by plants provided a source of food that could be exploited by animals, but in order to exploit it, animals had to be able to breathe atmospheric oxygen directly, rather than obtaining it from the oxygen dissolved in water, and they had to retain water in their bodies sufficiently well to maintain the fluid environment in which materials are moved about the body and in which most biochemical reactions take place. Some fish were able to obtain oxygen from the air in their swim bladders and adult amphibians, reptiles, birds, and mammals possess lungs, although amphibians obtain much of their oxygen through their skin. Insects, which were the first animals to move on to dry land, breathe by a mechanism involving diffusion, a process that is very efficient over short distances but which would be inadequate for any animal larger than an insect. Although amphibians also obtain oxygen by diffusion in their adult forms, they augment this by breathing with lungs. Dehydration remained a hazard for fish that left the water, and for amphibians, because their skins are not entirely watertight. But reptiles have watertight skins, and the possession of a watertight skin and the production of amniotic eggs enabled them to move far from water.

Reptiles spread out to colonize most of the world's surface. Their fossils have been found in all the continents except for Antarctica, and it is likely that there the fossils lie beneath the ice sheet. It is believed that at one time all the dry land on Earth comprised to form a single continent, Pangaea, which by the Cretaceous had divided again to form Gondwanaland in the south and Laurasia in the north. Then Gondwanaland subdivided, losing first India and then the combined land mass of Australia and Antarctica, and much later breaking into what are now Africa, Madagascar, and South America. Laurasia was split as the North Atlantic opened across it, but it did not break into its present masses until much more recently. Reptiles appeared first during the Pennsylvanian Period of the Paleozoic Era, perhaps 300 million years ago. By the time Pangaea broke into two they were established on land and were continuing to evolve into new forms.

This evolution led to the emergence of the archosaurs—true to its Victorian origins, the word means "ruling reptiles"—and so to the dinosaurs proper. The snakes and lizards, most typical of modern reptiles, appeared later, the lizards during the Triassic and the snakes during the Cretaceous. The only archosaurs to survive to the present day are the crocodilians.

The evolution of the dinosaurs is complicated by the fact that many of them continued to evolve in parallel after they had diverged from the ancestral stock, so that there are similarities among groups that are related to one another only distantly. The first definitely protodinosaurian group to appear, and so the ancestor of all later groups, was that of the thecodonts. They survived the major extinction at the end of the Permian Period, some 225 million years ago.

The thecodonts and all the early archosaurs were carnivores, which fed upon amphibians and earlier reptiles. They were able to hunt other animals because they could walk faster than their prey. They were descended from individuals in which the pelvis differed from the amphibian and earlier reptilian form in one of two ways—either their limbs were brought into a position directly below the body (in amphibians and such animals as crocodiles the humerus and femur are mounted horizontally and only the lower limbs are vertical) or they permitted bipedalism.

Most of the archosaurs were bipedal, although some were only partly so, and later in their history some formerly bipedal species abandoned bipedalism. Those that did so betrayed the route by which they had reached quadripedalism by having hind legs that

were longer and stronger than their forelegs and long, strong tails
that had helped the bipedal forms to balance.

As carnivores, the first archosaurs had pointed teeth appropriate
to their diet, but some later archosaurs had teeth which suggest a
vegetarian diet, and there were some that lost their teeth entirely
and had hard, horny bills. There were other physiological characteristics, relating to the skull as well as to the limbs, by which
archosaurs were distinguished from other groups of reptiles.

It was from among the archosaurs that the dinosaurs sprang. We
tend to group all the dinosaurs together, but they comprise several
distinct groups. While some continued to live as carnivores, by the
time they can be identified as dinosaurs, most of them had become
vegetarians and walked on four legs rather than two. The archosaurs gave rise to the Saurischia and Ornithischia, crocodiles,
pterosaurs, and birds. Each group is distinct, and although birds
are not reptiles, they are included because they diverged early from
the reptilian stock, lived during the period when the reptiles were
the most numerous of large land vertebrates, and retain many
reptilian characteristics. Oddly enough, they are not related closely
to the pterosaurs, the winged reptiles, which evolved by a different
route. The two orders of the dinosaurs proper are the Saurischia
and the Ornithischia.

The groups differ in the structure of the pelvis, but they differ in
other ways as well. The saurischians began as carnivores, and while
many of them evolved into four-footed herbivores, all the carnivorous dinosaurs belong to this group. The ornithischians were herbivorous from the start and remained so, but like the saurischians,
some developed into large animals and many were bipedal, at least
partly. Some of these two-legged forms evolved into the hadrosaurs, the bizarre duck-billed dinosaurs believed to have been grazing and browsing in large herds at the very end of the Cretaceous.

Within the two main groups there were many lesser groups and,
within them, innumerable genera and species. They came in all
shapes and sizes and not all of them were large. Some dinosaurs
were no bigger than a goose. Some of them acquired bony plating,
possibly for defense, or great bony frills which in some species
(such as *Protocerotops*) seem to have been formed from a pulling
back of part of the skull to provide more room for jaw muscles.
The dinosaurs were obviously diverse and distributed widely, and
we may suppose that they exploited all the resources available to

them. Most definitely, at the end of the Cretaceous they were not in decline. In the words of A. S. Romer (*Vertebrate Paleontology*), "Except for the primitive paleopods and the stegosaurs, all of the major types of dinosaurs were still in existence in the closing phases of the Mesozoic [the Cretaceous Period]; but by the beginning of the Cenozoic [the Tertiary Period] all had vanished."

The dinosaurs died as the result of some kind of catastrophe. Had the catastrophe not occurred, would they have become extinct anyway at about that time? As mammals ourselves, we are acutely aware of the advantages that accrue from our two principal features: homoiothermy and viviparity. Descended from "mammallike" reptiles, mammals were present during the later part of the Cretaceous and stable systems can be destabilized if new species invade them and begin to establish themselves. In our own time ecological changes have been wrought by the introduction of outside species into previously stable ecosystems—rabbits, feral goats, and feral cats have caused such changes in many parts of the world.

In such cases, of course, the new species are always strangers to the system they invade. Had they evolved within the system they would have formed part of a network of relationships that included predators and parasites, and their numbers would have been checked—a stable system cannot be disrupted in that way, from within. Could the mammals have invaded? Is it possible that they, or at least some of them, appeared first in a particular area, that they migrated from that area, and that their relationship with the dinosaurs was comparable to the relationship between, say, the Australian marsupial mammals and the placentals introduced by Europeans into Australia, which took over much of their habitat and caused the extinction of some of them? Yes, it is possible, although there is no evidence that would lead us to suppose that is what happened, but the explanation is insufficient. The placental mammals displaced the marsupials in Australia certainly, but only some of the marsupials became extinct, not all of them, and the largest marsupial herbivores, the kangaroos, may even have benefited and become more numerous.

Should we suppose, then, that the example of marsupials versus placentals is insufficiently extreme? Should we ascribe a much greater advantage to the mammals in their competition with the reptiles? That is one traditional view, but how real may that advantage have been?

Mammals are homoiothermic and viviparous, as we have said. Both traits confer some advantages, although homoiothermy must be paid for in a much greater food requirement. The sudden appearance of homoiotherms might well strain food resources that previously had been barely sufficient for the more modest appetites of the poikilotherms. No doubt some reptiles would have been living at the edge of their range, amid resources that were sparse. In such circumstances, however, poikilotherms would have been unlikely to have been displaced by homoiotherms. Where food is scarce, the advantage does not lie with the glutton! Provided the environmental conditions remain stable in a habitat where poikilotherms thrive, homoiothermy offers no advantage. The advantage is realized when the homoiotherms migrate into environments from which poikilotherms are excluded by climate. It is entirely possible to imagine a situation in which homoiotherms are able to exploit resources for which the preexisting poikilotherms have no use, so that mammals live alongside reptiles albeit in a subordinate position as it were. But it is difficult to see how they may gain the upper hand unless and until some outside event changes the environment to the disadvantage of the poikilotherms.

What is more, we are assuming that the reptiles *were* poikilotherms. As we suggested earlier, they may not have been. The mammals had evolved but recently from those mammallike reptiles, which may well have been homoiothermic, and homoiothermic ancestors of birds may also have been alive. Even if we allow an advantage to the homoiotherms, it is an advantage the early mammals are likely to have shared with at least some of the reptiles.

Viviparity, too, confers an advantage, but even here the situation is no more helpful to the mammals than to some reptiles. Many reptiles today are ovoviviparous, as we said above, which permits them to enjoy most of the same advantage. And since viviparous mammals descended from reptiles, it is not impossible that in the Cretaceous some of the reptiles were also truly viviparous.

In a sense, then, we are not necessarily comparing like with like. Many reptiles are ovoviviparous and some may have been viviparous. Some reptiles may have been homoiothermic. But we have no reason for supposing that many, or even any, reptiles were both viviparous and homoiothermic—as all mammals are.

If we are to imagine a catastrophic situation in which animals possessing both traits might thrive at the expense of animals possessing only one or neither of them, we need to introduce some kind of environmental change brought about by external factors. The most likely candidate for such a factor is the climate. If the climate had become more markedly seasonal, the reptiles might have suffered.

The summers would need to be no cooler, provided they were warmer than the winters and that winter temperatures fell low enough to inhibit poikilotherms. If the summers became warmer this, too, would have had an adverse effect on poikilotherms. Whenever the temperature was higher or lower than the range within which they could function, they would have had to devote to warming or cooling themselves time that could not be spent in other activity.

If these animals laid fertilized eggs, the eggs, too, would have been vulnerable to high and low temperatures. It may be that animals that evolved in a climate without distinct seasons would not have bred seasonally. There would be no reason for them to do so. Once the climate became seasonal, selection pressure would favor a change to seasonal breeding, but this might have been insufficient by itself to ensure survival. It is not only the metabolism of the animals which is affected by temperature; so is the vegetation on which they depend for food. A seasonal climate is one in which the availability of food varies seasonally, too.

This would have added a further selection pressure on breeding behavior. It is not enough for an animal to lay its eggs when temperatures are right for incubation—the hatchling must emerge into an environment in which food is plentiful. Unless it can grow very rapidly to its adult size—which limits the strategy to small animals —this means it has to hatch early in the plant-growing season in order to have as long a time as possible to feed and gain weight in anticipation of the food shortages the following winter. However, if the eggs are to hatch early in the year, they have to be laid even earlier—and there will be an increased risk of cold weather chilling the eggs and killing the embryos.

There are two ways to solve the problem. Birds that live in high latitudes, for example, lay their eggs early in the year but incubate them with the warmth of their own bodies. But they could not do this were they not homoiotherms—it cannot be done by modern reptiles.

The second way is viviparity or, at the very least, ovoviviparity.

By either of these means, the environment for the incubating eggs can be controlled constantly by the adult. As it seeks the conditions it prefers, automatically it seeks conditions that favor its eggs. The strategy works for both poikilotherms and homoiotherms, for as the adult poikilotherm warms or cools its own body, so it warms or cools the eggs it carries. Nevertheless, the homoiotherm enjoys a clear advantage because it is less constrained by outside temperature and can function efficiently within a larger temperature range. It can conceive in autumn, gestate through the winter, produce offspring as winter draws to its close, and have them weaned by the time plant food is plentiful.

The viviparous method is modified in some modern mammals to permit the resorption of fetuses that are not viable, so ensuring both that the individuals that are born are those with the greatest chance of survival and that the resources the mother has invested in her fetuses are not wasted. In some mammals, implantation of the fertilized ovum is delayed, sometimes for several months. This permits the adults to mate at the time of year when they are at their most active and the young to be born when conditions for them are at their best. Neither fetus resorption nor delayed implantation are possible for an animal that is not viviparous.

So mammals and birds will thrive in markedly seasonal climates, while the great majority of reptiles thrive in nonseasonal climates. If the climate changed during the Cretaceous, it might have allowed the mammals to replace the reptiles, and if the climate changed suddenly, the change too might have been sudden. Did the climate change in this way?

During the Cretaceous Period, climates were benign all over the world and, which is more important, they were stable. They provided conditions that were ideal for the reptiles for the simple reason that they were the very conditions into which the reptiles themselves had evolved. Had that climate remained unchanged until the present day, there is no reason to suppose that the reptiles would have declined. Mammals might still be living in nocturnal obscurity and had "intelligent" beings evolved—let us say beings with advanced technologies—they might well have scaly skins and (probably) long tails.

However, the climate was changing at the end of the Mesozoic Era and the beginning of the change can be detected in the Cretaceous. By the end of the period the change had not proceeded very far—there was no ice on the surface—but it was growing cooler. Mountain building had altered air circulation patterns and changed local climates. Crustal movements had caused continental land masses to be raised and sea levels to fall. Perhaps changes in the Sun itself had caused a reduction in the amount of energy reaching the Earth. These changes may have been enough to cause local extinctions, and there is fossil evidence (see J. David Archibald in *Nature*, Vol. 291, p. 650, 1981) that in North America such changes did cause extinctions, but there is no reason to suppose they were more than local.

The climatic change continued into the Tertiary Period, sea levels rose again because of further crustal movements, but even so, the cooling was slow. Until the Oligocene Epoch, which began about 38 million years ago, tropical climates extended to about 50 degrees either side of the equator, and it was not until about 27 million years ago that climates over much of the Earth became markedly seasonal and the tropical belt shrank toward the equator. By that time, the advantages of combined homoiothermy and viviparity might possibly have begun to tell and the mammals might well have been multiplying while the reptiles declined.

But that, of course, was not why the dinosaurs disappeared. They disappeared because a catastrophic event took place that removed large animals. After the event, conditions rapidly returned to normal and the mammals proliferated to fill the ecological niches vacated by the reptiles. And the dinosaurs vanished 65 million years ago, not 27 million years ago, about 38 million years earlier than they would have if the principal cause of their disappearance had been climate change. Had the impact event not occurred, certainly the climatic cooling would have produced evolutionary changes as species adapted to the new conditions. We can only assume, however, that in such a case the mammals would have profited. For all we know, reptilian species might have appeared that were better equipped for the new circumstances under which they were living. To some extent, though, this is merely to play with words. Is not a homoiothermic, viviparous reptile, adapted to life in a seasonal climate, but a mammal by another name?

Yet there is an interesting evolutionary consequence, even so. The great success of the mammals began about 65 million years ago, in the time immediately following the impact. Had the impact not occurred, the most likely cause of major evolutionary change *would* have been the climate change, about 38 million years later. It looks, therefore, very much as though the principal long-term biological effect of the impact was to bring about a change some 38 million years earlier than it might have taken place otherwise. The impact accelerated evolution by this amount. Depending on the view we take, either we have the impact to thank for the existence of human beings at all or, if we assume the mammalian proliferation would have happened anyway, we must thank the impact for the fact that we are here now. Without it we would not have made our entrance for another 35 million years or more and today the first true hominids would not yet have appeared although there might well be some slightly manlike apes among the primates.

It is usual to think of a major catastrophe in terms of the devastation it causes and the number of living organisms it kills. This is only one, short-term, way to view the effects, however. When we look at the long-term consequences, we have to admit that the effects were insignificant—life continued. And being human, we regard them as beneficial—there is reason to suppose that in this case a catastrophe has given us some tens of millions of years tenancy of our planet that otherwise we would not have enjoyed. It does not follow that all major catastrophes will have such an effect —probably only those that occur during long periods of stability exert such leverage. A catastrophe occurring during a period of rapid change is likely to be much less beneficial. Just because this particular catastrophe accelerated evolution—if it did—does not mean that on behalf of the planet, and far less on our own behalf, of course, we should welcome the prospect of further catastrophes. If there were another, the consequences might be very different. In this case, though, it does look as though no great harm was done in the end.

10

The Violent Sky

We now believe that 65 million years ago a planetesimal, or less probably a comet, collided with our planet. It killed countless billions of living organisms and changed the course of evolution. We cannot help asking the questions that spring of themselves from this discovery. Was the event an isolated one? Had something like it happened before? Will it happen again?

The place to begin our search for answers is the Moon, and when we have examined our own satellite, we should compare our findings with evidence obtained from our near neighbors in the solar system, Mars, Venus, and Mercury.

The first thing to note about the Moon is that it is heavily pockmarked with craters. For a long time the origin of these craters was uncertain. It was thought that they might have been formed by impacting bodies or, on the other hand, have originated volcanically. When close examination of them became possible, however, it was determined that while some were formed by volcanoes, the great majority were impact craters.

Almost all of them are circular; this is curious because, as we saw earlier, the chance that an impacing body will strike another body's surface precisely at right angles is very remote, and the chance that all impacting bodies would do so can be dismissed entirely. If a body were to strike at a shallow angle, however, surely it would form a crater with a more oval shape; circular craters are more likely to be caused by volcanoes. But the riddle was solved, and its solution is relevant to our story.

When a meteorite weighing many tons strikes the surface of the Earth or Moon at a speed measured in kilometers per second, it transfers its vast momentum to the rocks at the surface. At first these are compressed and displaced downward, but soon the ever-increasing pressure of the solid rocks below resists the impact force and halts the downward motion. Most of the impact energy now resides in a small region of hot, highly compressed rock, which expands radially outward and thrusts the surface rocks aside to form an almost perfectly circular crater. By the time this radial outflow starts, the forward (downward) motion of the meteorite will have all but ceased. Consequently, craters are circular whatever the angle of impact.

In the first fierce compression, as the meteorite is slowed by the impact, enough heat may be generated to turn it into a gaseous plasma. When this expands to form the crater, little or no solid trace of the original matter that fell from the sky remains on or below the crater floor. It is dispersed as dust and carried away by the winds.

Small meteorites are often slowed sufficiently by the Earth's atmosphere—like reentry vehicles carrying astronauts—to land comparatively gently and so to remain intact. Large or planetesimal-sized objects lose so small a proportion of their total kinetic energy in the air that the presence of the atmosphere has a negligible effect on the force of their impact at the surface.

The Moon, then, has been struck many times in its history by large, solid bodies moving at high speed. When unmanned spacecraft sent back the first close-up pictures of the surface of Mercury, these also revealed a heavily cratered surface. Our information about the surface of Venus is less detailed, but Mars was found to be heavily cratered which leads us to suppose that Venus must have suffered in the same way. In fact, all the bodies in the inner solar system—Mercury, Venus, the Moon, and Mars—have been hit countless numbers of times. Why should we suppose the Earth to be specially privileged? Why should objects strike the Moon but not the Earth, which is very close by and presents a much bigger target?

There can be only one answer. The Earth is not especially favored. It, too, has been struck by such planetesimal bodies many times. Its surface retains few craters simply because it is tectonically active and because its abundant water causes weathering. Given sufficient time, entire mountain chains are eroded by the Earth's weather. It is hardly likely that craters could survive for long. The surface of the Earth retains impressions poorly, and it may be that the surface of Venus, which appears to be only sparsely cratered, also retains them poorly. In its case, perhaps the severe erosion is caused not by water but by an atmosphere so dense that it might be thought of as a vast ocean of carbon dioxide. The surfaces of the other nearby planets retain craters well.

This answers the first and second questions. The planetesimal impact we have described was not an isolated event. Such impacts had undoubtedly happened before, many times. This one occurred 65 million years ago, and that seems a long time. Does it mean that the last impact was the final one, that the sky contains no more large objects with which to bombard us? Alas, the answer to this question must be no.

Our lives are short and we measure time in relation to them. To our minds there is little difference between a period of a few million years and eternity. Such large numbers are beyond our experience. We must think of events within the solar system in relation to what we may call "solar time," not "human time." A human life span may, rarely, extend to a hundred years. The life span of the solar system extends for thousands of millions of years. It has existed already for some 5 billion years. Let us relate the two, by converting solar time into human time. To do this we must make an assumption about the life span of the Sun and give it an "age." Let us suppose it to be middle-aged, in human terms about fifty years old. This means that fifty human years are equivalent to 5 billion solar years, and so what is 65 million years in the solar system in human terms would amount to rather less than eight months. Therefore an event that occurred eight months ago and which is known to have occurred many times in the past is clearly a recurring event. That the impact took place long ago is an illusion.

How frequently do such impacts occur? This has been calculated on the assumption that the Earth experiences impacts as frequently as do the Moon and other inner planets and because wherever craters can be examined closely they can be dated by the rock strata around them. The picture that emerges is that early in the history of the solar system impacts were very frequent. Then, around 3 to 3.5 billion years ago, the frequency fell sharply to a level that has been maintained ever since. Thus many of the craters we see on other planets may be very ancient—although ancient craters are liable to be overlain by more recent ones. The present frequency has been estimated by W. M. Napier and S. V. M. Clube, of the Royal

Observatory, Edinburgh (*Nature*, Vol. 282, p. 455, 1979), who calculate that a planetesimal with a diameter of about 3.9 kilometers (2.4 miles)—the number calculated from the size of observable craters—will collide with the Earth about every 9.2 million years, one with a diameter of 5 kilometers (3 miles) every 14 million years, one with a diameter of 10.9 kilometers (6.8 miles) every 58 million years, and one with a diameter of 30.8 kilometers (19.1 miles) every 360 million years. The meaning of this is obvious. The next impact of a body the size of that which caused the Cretaceous extinctions is already overdue! There is some small consolation for the more nervous among us: other scientists have made these calculations and some of them estimate the frequency of impacts from bodies 10 to 11 kilometers (6 to 7 miles) in diameter to be about every 100 million years.

We have found answers to the three questions we asked at the beginning of this chapter. Large impacts have occurred in the past, indeed they are rather commonplace events, and they will occur again. The fact that they occur with a frequency that can be measured leads to a further thought. If it was an impact that brought the Cretaceous to an end, did other impacts also mark the end of other geological periods? In at least some cases, this is a distinct possibility and several scientists have suggested it in recent years. Napier and Clube found evidence of impacts whose dates correspond fairly closely with the commencement of the Pleistocene (1 million years ago), of the Pliocene (13 million years ago), of the Miocene (25 million years ago), of the Oligocene (36 million years ago), of the Eocene (58 million years ago), of the Jurassic (181 million years ago), and, though with less certainty because of difficulties with precise dating, of the Triassic (230 million years ago) and the Carboniferous (345 million years ago). They suspect a similar event in connection with the Permian (280 million years ago), the commencement of the Devonian (405 million years ago), and for both the beginning and end of the Cretaceous.

A difficulty may have occurred to you. If indeed our history is punctuated by large bangs as the planetesimals arrive, why does the solar system not exhaust its stock of suitable planetesimals? After all, each body can strike a planet only once, and so each strike means there is one less body available. Yet the frequency of the impacts can be calculated and it seems to be fairly constant. At all events there is no sign of its decreasing.

This is a genuine difficulty and it has been noted by people who have devoted much thought to it. The question has not been answered, but there are several theories. In the first place, we must try to account for the sharp drop in the frequency of impacts since about 3.5 billion years ago.

The solar system is thought to have condensed from a nebula of gas and solid debris. The exact mechanism by which the planets aggregated is still debated, but among the more favored theories is that of aggregation through the serial impact of planetesimals. According to this theory, colliding bodies adhered to one another, and as some of the aggregations grew large, the smaller rocks that struck them added to their mass, just as impacting meteorites do today. Once the present family of planets became established, the planetesimals that remained would be depleted progressively until the rain of impacts all but ceased. The date of the peaceful conclusion marking the assembly of the solar system was just before 3.5 billion years ago.

The frequency of impacts fell, but why did it not fall all the way to zero? If each spanking new planet had swept up such material as was available to it, any remaining lumps must have been in secure orbits that did not bring them close enough to a planet for a collision to occur. We might allow that some of these orbits were very large, so the period of the small bodies (the time that elapses between each of its passages by a given point in the solar system) would be large, but even so, they should all have disappeared long ago. The most typical debris remaining from the formation of the solar system is to be found in the asteroid belt where, some astronomers have suggested, a planet was prevented from forming because of the gravitational influence of Jupiter, which tore the lumps apart whenever they accreted to more than a certain size. However, the asteroids have known orbits and it is difficult to see how they might wander from those orbits in our direction or in the direction of any other planet.

There may be material that still remains from the formation of the system and perhaps the planets, including the Earth, continue to collect it. The aggregation of small particles into larger lumps was more probable when the cloud of material was relatively dense. Once most of it had been accreted to form the Sun and planets—and more than 99 percent of the total mass of the solar system is accounted for by the Sun itself—what remained was too diffuse to accrete into larger bodies. Some of this remnant may even fall in the rain of small particles we see as meteors—"shooting stars." The Earth collects as much as 1,000 tons of them a year. There are also the comets, which may or may not originate within the solar system, and the Apollo objects, which may or may not have been comets. So far as we know, however, there is not enough material left over from the assembly of the solar system to account for major impacts on the scale of those examined by Napier and Clube.

It begins to look as though material is able to enter the solar system from outside. If so, this is an exciting prospect, because it means we may be able to sample material that did not originate within the solar system. It could tell us a great deal about other regions of our galaxy.

The first possibility for an extrasolar source was proposed in 1950 by the eminent Dutch astronomer Jan Hendrik Oort. It developed from earlier ideas by which astronomers had sought to explain certain anomalies in theories of the formation of the solar system. Essentially these concerned the velocities and angular momenta (the mass multiplied by the velocity) of the planets. If the planets formed around the edge of the mass that was to become the Sun, how could they move faster than the rate at which the Sun rotates? Jupiter has more than 30 times and Neptune more than 80 times the angular momentum they would have if that were the whole story of the formation of the solar system. If some passing star were to accelerate the planets to their present velocities, why did they not escape from the Sun altogether? The difficulty resolved itself when the Sun was made one—the survivor—of a pair of stars that once formed a binary system. Binary star systems are so common in our galaxy as to be the rule rather than the exception. If some mechanism—perhaps that encounter with another star, although there are other hypotheses—destroyed the Sun's companion, the behavior of the planets can be explained.

What happened to the matter from which the companion was composed? Oort suggested that it, or some of it, remained to form a cloud of small bodies, rotating about its own axis and orbiting the Sun, but lying invisible beyond the orbit of Pluto. He argued that this cloud—known today as the "Oort cloud"—is the source of the comets and of all other planetesimals that move in orbits markedly different from those of the planets and asteroids.

Strictly speaking, the Oort cloud must be composed of the same material as the solar system, of which it forms part, and so bodies that enter the inner solar system from it cannot be considered as extrasolar. However, there is an alternative theory, which probably is incompatible with the existence of the Oort cloud.

Although it appears that the Earth experiences large impact events at intervals that can be estimated, this does not mean they occur as discrete single events separated by intervals of time. They "bunch"—that is to say, one large impact may be accompanied by many smaller impacts at about the same time—within tens of thousands or at most 1 or 2 million years of the major event. This raises the possibility that from time to time the solar system encounters clouds of matter consisting of particles ranging in size from the minute to the planetesimal. Some of these particles may thus enter the solar system and some collide with bodies within the solar system.

We know that the planets orbit about the Sun; this is easy to observe. We know, too, that the planets and the Sun all rotate about their own axes. This, too, is self-evident in the case of Earth and not difficult to observe in the case of other planets. It may be more difficult to appreciate that the solar system as a whole is involved in movements on a larger scale.

Our galaxy, which consists of a relatively dense, roughly discus-shaped center with arms spiraling out from it, also turns on its own axis, and as a component of the galaxy, the solar system turns with it. The motion is complex, because the galaxy is not a solid body, and different parts of it move at different speeds. We can calculate, though, that since it first formed, the solar system has made about twenty-five complete rotations about the galactic center—that is, it is twenty-five galactic years old. In addition to this, the Sun, like all stars, has its own more irregular movement which carries it to different regions of the galaxy. This is not easy to understand, but an analogy devised by I. S. Shklovskii and Carl Sagan (see *Intelligent Life in the Universe,* by I. S. Shklovskii and Carl Sagan, San Francisco: Holden-Day, 1966) may help. They invite their readers to picture the galaxy as though it were a cloud of gas in which the stars represent molecules. The first thing to note is that the cloud is very rarefied indeed, but the molecules in it are not distributed evenly: the cloud is denser in some places than in others. The whole cloud is rotating and so its molecules are engaged in that rotation, but like real molecules in a real gas cloud, they also wander about in response to more local forces.

A similar movement carries the Sun through the spiral arms of the galaxy and we cross such an arm every few tens of millions of years. If these spiral arms were to contain planetesimals, it is possible that the movement of the Sun through them would cause some of them to be caught to form an Oort cloud. The cloud would not endure for very long (in galactic terms) because its own movement and the gravitational forces exerted by the Sun would cause its dissipation, partly by throwing bodies out from it and partly by drawing them closer to the Sun itself—into the solar system—where many would be lost by encounters with planets. Napier and Clube, who support this theory, suggest that most comets, planetesimals outside the asteroid belt, and many satellites were captured in this way, and that their residence within the solar system is temporary. Eventually they will be lost, either by collision with a planet or by moving outside the system entirely.

There are difficulties with the theory, which Napier and Clube acknowledge. In particular, it is not easy to account for the formation by accretion of a requisite number of large bodies in the spiral arms in the time available, as judged by estimates of the age of the galaxy as a whole. Nevertheless, such formation is not impossible, and the theory is persuasive.

If we are sampling extrastellar material as we journey through the galaxy, how does this square with the argument that since analysis of meteorites shows their composition to be typical of that calculated for the solar system, they must have originated within the solar system?

We know that the fundamental material of the universe is hydrogen and that all heavier elements are made from it. We know how they are formed, within stars, and we know, too, that elements heavier than iron must be formed in conditions produced only in supernovae. We have theories to explain the formation of our own solar system, whose composition is known. It is not possible that our system, or even our region of the galaxy, is unique. The laws that govern the formation of elements and the formation of stars must have universal application, at least within our own galaxy. Consequently, material sampled anywhere within the galaxy may well be composed of elements distributed much as we find them locally, since their formation and their arrangement into solid bodies

has resulted from processes identical to those which operate locally. The idea that planetesimals are formed by accretion of dust
within the spiral arms of the galaxy is consistent with theory. It is
also consistent with observation, which shows a depletion of carbon, nitrogen, and oxygen in the interstellar dust and the existence
of "cold globules" that emit carbon monoxide, possibly associated
with planetesimal formation. The globules appear to be accretions
of dust that probably are forming and fragmenting rather in the
manner by which stars form in their very early stages. The fact that
they are observed to emit carbon monoxide, which can be detected,
leads Napier and Clube to infer the existence of such globules in
parts of the cloud where carbon monoxide emissions occur but
where solid bodies cannot be seen.

The distribution of isotopes of elements may be typical of the
particular stellar furnaces—supernovae—in which they formed, but
there is no real reason to suppose that this is so. Until we discover
samples of elements whose isotopic compositions differ markedly
from one to another, we cannot assume that the stars in which
heavy elements form impose their individual signatures upon those
elements.

In the end, we do not know whether the Cretaceous planetesimal
originated within the solar system or whether it was captured, but
the possibility that planetesimals do enter the solar system from
time to time is strong. What we do know is that planetesimal impacts have occurred many times in the past, that for the last 3 to
3.5 billion years the frequency of impacts has remained fairly constant, and that if there does exist a finite stock of impacting planetesimals, it does not seem to be exhausted. One day, more probably
sooner than later, given the known average frequency of major impacts, one will strike us again.

What can we do about it? In theory, two approaches are open to
us: prevention or cure.

In reality, the "cure" approach would be impracticable. It would
require us to accept the inevitability of the event, but to seek to
survive it. Not all of us could survive it, of course. In the region of
the impact itself all life would be destroyed. In more remote regions, however, survival might be possible provided precautions
were taken in advance. Most obviously these would necessitate the
building of shelters, which would need to be capable of filtering
and purifying incoming air. They would need to withstand consid-

erable heat and shock while maintaining habitable conditions inside, and they might need to withstand total immersion in seawater. The people living in them would have to remain in them for several weeks, so they would need adequate stocks of food and water. In principle, such requirements present no great technical difficulties; we can construct buildings to these kinds of specifications.

We would have to allow for the likelihood that when people emerged, they would find that, for one reason or another, all large animals had been killed and, probably, that most plants had died. The soil might become very acid and water contaminated. We can allow for this. If food stocks are adequate to last for at least two years and if a breeding nucleus of farm animals is taken into the shelters, it will be possible for animal husbandry to recover. If seeds for farm crops and also for those plants that grow naturally in the immediate area are taken into the shelters, the vegetative cover can be restored. If ample supplies of lime are stored, the acidity of the soil can be corrected. If some of the shelter buildings house not people but water purification plants, a supply of potable water can be restored quickly. It can be done, all of it, but is it feasible?

In the first place, who is to be saved? Unless we suppose the entire human population is to be accommodated in our shelters, we will face some difficult moral decisions, and the most likely— perhaps the most sensible—political response to them may be to make no decision at all. The cost of the preparations will be vast, and much will have to be met by communities that so far have failed to provide adequate food and shelter for millions of human beings in normal times. The expenditure will have to be justified against some evaluation of the risk to which people are to be exposed. This would require us to estimate not only the magnitude of the effects of the impact, but also its probability. We can say with some confidence that an impact of this kind is very probable within the next 50 million years, but that is hardly likely to impress an accountant! He will want to know the probability within the next ten years, or five. If there is a major impact every 100 million years, the chance that one will occur within any period of ten years is 1 in 10 million. Of course, the odds are the same for any ten-year period, including the ten-year period in which the event happens. But that does not help us to identify the fatal decade.

We may assume, therefore, that no preparations will be made to survive the event. Can it be prevented?

This is a good deal more practicable, extraordinary though it may sound. Indeed, a few years ago the science fiction film *Meteor* (1979) postulated just such an event, although it obtained its planetesimal improbably—from the asteroid belt. The theory is sound enough, however. Provided the body can be detected while it is still a long way from Earth, it should be possible to fire missiles at it that are designed to explode in its immediate vicinity—not necessarily on its surface—and will deflect it. The task sounds prodigiously difficult, but it might not be so. The missiles would be guided to their target, and they are not required to score a direct hit. If they explode nearby with sufficient energy, the body will be deflected, and it matters little in which direction it is deflected since any deflection would carry it away from Earth.

How much energy would be needed to deflect it? That depends on the amount of deflection required, which depends in turn on the distance the body is from the Earth when it is deflected. If you remember our problem of firing a missile at the Moon, you will recall that the angle between one edge of the target and the other is very small. Imagine now that instead of firing at the Moon you are firing at Mars. This is a very much larger body, but because of its greater distance from us it appears as a mere point of light in the sky, not much different from a star. The angle between one edge of it and the other is a tiny fraction of 1 degree and that angle diminishes as our distance from the target increases. If we could detect the planetesimal when it was still many millions of kilometers from us, if we could compute its trajectory and so confirm that it was on a collision course, if we had missiles available that could be launched and then guided in flight from Earth (existing military long-range missiles are programmed in advance and cannot be redirected in flight) so that we could intercept the planetesimal while it was still some millions of kilometers away from us, then a few dozen megatons of energy might be sufficient.

Might it be practicable to construct a defensive system of this kind? Again we must weigh the risk and the cost of averting that risk. The odds against an impact this year, or next, do not change, but we might be more interested in the consequences of such an impact, because the cost of the defense system would be very small. It would call for a modest stock of missiles with warheads, which exist already in superabundance, and the major powers could put aside an agreed number each for this purpose with no loss to their military capabilities. Already there are plans to build orbiting observatories, and if the routine scanning of the solar system is not one of the tasks they are designed to perform—and almost certainly it is—then it would cost little to add such a task and the equipment necessary for it. The missile guidance systems may not exist as such, but it would not be difficult for organizations capable of landing small vehicles safely on the surface of Mars and Venus to build them. It could be done and should be done. The cost would be little, but the consequences of a major impact would be almost incalculable.

For those among us who favor a return to a simpler way of life, less dependent on advanced technologies, as a route to human survival, there is a lesson in all this. The event which could wreak havoc on a global scale more certainly than any other can be averted, but only if we devote the most advanced of our technologies to the problem. If the risk is real—and it is—human survival and very possibly the survival of most life on this planet can be assured only with the aid of high technology. It is not that life itself might be extinguished on the planet, but that the direction of its future development might be changed. Our ancestors may have benefited from the last such change, but would we benefit now? Can we be sure there is not some small, insignificant group of animals whose further development awaits our departure? Next time will we fill the role of the dinosaurs?

Before we leave the planetesimal, perhaps we should try to learn more from the Earth's experience of it and of its fellows. How does this threat compare with others of which we hear more?

11

Survival

Throughout this book we have discussed a single catastrophic event. Since the 1960s we have grown accustomed to prognostications of future catastrophic events that might be engineered by human beings themselves. Most of these have been discredited. Why, then, should we ask you to accept the reality of an event that seems, on the face of it, less probable and a great deal more remote in time?

To begin with, perhaps we should consider the most disastrous event most of us can imagine: full-scale thermonuclear war. This, it is often said, would spell extinction for the human species and possibly the end of all life on Earth. The planet would be burned to a cinder.

Not surprisingly, the likely consequences of such a war have been calculated—in fact, by a group working under the auspices of the United States National Academy of Sciences. They supposed a full-scale war between the NATO and Warsaw Pact forces, in which all available weapons were used. Within the territories of the combatant countries themselves devastation would be widespread and the death toll would run into many millions. Outside the combatant countries, and most especially in the southern hemisphere—this would be a northern-hemisphere, rather than global, war, after all—there would be very little effect at all, mainly because of the very limited extent to which air is exchanged across the equator. Within thirty years all affected natural ecosystems would have recovered and the Earth would sustain no injury that would leave permanent scars. The war would be traumatic as far as humans were concerned, but there is no possibility of the extinction of our species and far less of the extinction of all life.

This is the finding we might have expected since we have access to some of the information on which it was based. We know that in the case of the two nuclear weapons that were used against Japan in 1945 human deaths were caused by heat, by blast, by radiation —mainly radiation emitted in the flash of the burst—by the infection of injuries that could not be treated for want of medical services, and from disease caused by the contamination of water supplies. Most of the deaths—some 125,000—occurred within a short time of the attack. Many people received high doses of ionizing radiation and many died from it, and it was expected that those whose exposure was less than immediately lethal and those who were irradiated mainly by the fallout some distance from the devastated areas would suffer later in life from diseases caused by the radiation. Some people did. But the types of disease—chiefly cancers—we associate with radiation may be caused in other ways and all humans are exposed constantly to natural radiation. Since the diseases occur anyway in human societies everywhere, we may attribute to radiation emitted by man-made devices only those cases that are in excess of the average incidence we would expect. Today, thirty-eight years on, the incidence of such diseases in those two Japanese cities is, if anything, rather lower than in populations that have not been irradiated. The idea that the radiation from nuclear weapons is the major cause of death, or that it increases greatly the birth of grotesque mutations, is without any foundation. We are not attempting to justify or excuse what happened at Hiroshima and Nagasaki, but the fact remains that far from being uninhabitable piles of scorched rubble today, both cities are thriving, prosperous, and populous.

We also have detailed information about the history of the Bikini atoll and Eniwetok, the Pacific sites used for the testing of thermonuclear weapons. These and the Soviet test sites in Siberia are the most heavily irradiated places on Earth. During the test period all life on them was destroyed and their soils were scorched and blasted away, leaving the bare rock exposed. When testing ceased, however, recolonization began and today it is almost impossible to distinguish the sites ecologically from similar sites in that part of the world that were unaffected.

We see, then, that the planet is well able to survive the worst disaster we can imagine ourselves engineering. Why should it not? The planetesimal impact we have described was thousands of times worse, and the Earth survived that.

This may help us place in perspective the idea that human activity is capable of destroying life on Earth. When we examine the gloomy predictions in more detail, we find that each of them is based either on false data or on quite exaggerated evaluations of human power. We will so pollute the seas that they will be unable to sustain life! By our destruction of green plants and especially of marine phytoplankton, we will deplete the atmosphere of its oxygen! By squirting our aerosol cans, we will destroy the ozone layer and be fried in ultraviolet light! By building more and more nuclear power stations, we will destroy ourselves (by mechanisms that are not specified)! The list is a very long one and in ridiculing it, we do not intend to ridicule legitimate concern about the state of the terrestrial environment. This is a matter for our proper concern. With the knowledge we possess we have no excuse for rendering the planet a less pleasant or hospitable place than it need be for ourselves or for any other species. All we would do is to delete the more extravagant statements, and the first and foremost of these is that human activity is, or could be, life-threatening to the human species generally—although we could destroy cities and even nations, of course, together with most of the people living in them—or to life as a whole.

How can we be so sure? It is a subject to which both of us have given much thought over the years and we have reached a similar conclusion by different routes. This conclusion leads us to regard the Earth and the life it supports in what may seem a new way. Very evidently, life is extremely resilient, recovering rapidly from injuries, so that the conditions necessary for life are restored. They are restored, as they were established originally, by living organisms themselves, and the entire surface of the planet, from the depths of the oceans to the top of the atmosphere, has been shaped by biological processes. The arguments supporting this view are to be found in *Gaia: A New Look at Life on Earth* (J. E. Lovelock, Oxford: Oxford University Press, 1979) so we need not summarize them here. They account for the resilience.

They also place human activity in proper perspective. We know, for example, that from time to time the regions of the planet in the higher latitudes are buried beneath ice sheets during glaciations. Any land surface in a latitude higher than about 45 degrees north or south has been glaciated and will be again. Glaciation destroys all plant and animal life and probably most soil microorganisms as well, and it causes the severe scouring of the land surface. It is devastation on a scale that exceeds by far the puny attempts of industrialists or even of farmers. Since glaciations have no lasting adverse effect on life as a whole, we may be fairly sure that high-latitude regions are "expendable," that no matter what we do to their land surfaces, the consequences are unlikely to be serious.

We can also be confident that while our industrial expansion causes environmental problems, these problems are unlikely to prove serious so long as we generate information—the ability to predict the consequences of and therefore control the processes we initiate—at the same or a faster rate. At present the major growth lies in the area of the information industries. Increasingly, environmental problems are identified and diagnosed rapidly and, despite the political problems many of them present, usually appropriate action is taken.

Many of our environmental fears derive from the observation that human beings modify their environment to make it more hospitable for themselves and for their domesticated plant and animal species. Man-made habitat is contrasted with habitat that was not man-made, and habitat that was not man-made is held to be "natural," so that, by implication, man-made habitat is "unnatural." Thus the virgin forest is "natural" and the farm that adjoins it is not. This simple error in observation and in logic has caused us a great deal of trouble. If we are to distinguish accurately between real threats and imaginary ones, we must correct it.

All species, from the simplest microorganism to the largest plant or animal, modify their immediate surroundings. They cannot avoid doing so. Even so basic an activity as respiration alters the chemistry of the atmosphere, and in the early history of the Earth the entirely natural process of photosynthesis altered it drastically to liberate free gaseous oxygen, which was intensely poisonous to many of the organisms alive at the time. They were accustomed to an oxygen-free environment, after all. The release of large amounts of oxygen into the air was perhaps the most serious pollution incident in the history of the planet—yet not only did life survive, it thrived as it did following the Cretaceous planetesimal impact. We need not go back so far in time to see environments made by nonhuman living species, for all Earth environments were made in this way. Some species even modify their surroundings in ways we can recognize at once. Beavers can and do transform woodland into pasture. Large grazing animals turn forest into prairie. When humans change their own surroundings, therefore, they are doing nothing that is not done by every other living species.

The use of the word "natural" is no more than a semantic confusion arising from a word that has, and has had, many meanings. It has carried heavy moral overtones at least since St. Thomas Aquinas wrote about "natural law," by which he meant "God's law." Consciously or unconsciously, we have incorporated it into our judgment of human behavior, contrasting "natural"—or "brutish" or "animal"—behavior with "civilized" behavior. For the last two centuries or so, people have tended to favor one kind of behavior or the other, and more recently many environmentalists have espoused the view of Jean-Jacques Rousseau that in a "state of nature" man lives in peace with his "natural" surroundings. People have come to believe in the myth of the "noble savage"—and they have been supported in this by many amateur anthropologists—for all the world as though it were true. The fact is, of course, that the "noble savages" became ourselves and there is not the slightest reason to imagine our recent ancestors were any more "environmentally aware" than we are. Indeed, such evidence as exists suggests they were a good deal less so.

We must conclude, therefore, that our power to destroy the world, or even ourselves, is quite imaginary, a product of our own hubris. If we were to acquire such power—and we might if, for example, we developed cheap and inexhaustible sources of energy by which we could raise the atmospheric temperature until an irreversible greenhouse effect became established—it is most unlikely that we would be unable to control it. We would do well to make very sure that our capacity to control our industries and technologies is not eroded, but that requires of us nothing more than vigilance. We must view with the deepest suspicion those who would have us reject industry and technology in general and information technologies in particular. Such people make our demise more likely, not less so.

The credible threats must come from outside the Earth and the impact of a large planetesimal is the most immediate of them. The probability that one day it will happen is so high as to come close to certainty. Sixty-five million years ago one changed, or at least accelerated, the course of evolution. What would one do now? We who can guess at the consequences have a moral obligation to consider them seriously and to take such modest, inexpensive steps as we can to avert them, for we believe they might be averted. We owe that much to those who will follow us, human and nonhuman alike.

If we seek other threats, we will find them. We occupy a violent, dangerous piece of sky. But that is another story . . .

Index